D1226770

WORLD ENERGY: THE FACTS AND THE FUTURE

WORLD ENERGY: THE FACTS AND THE FUTURE

DON HEDLEY

Facts On File, Inc.
119 W.57th Street, New York, NY 10019

World Energy : The Facts and the Future

First published in the United Kingdom in 1981 by
Euromonitor Publications Limited,
18 Doughty Street, London WC1N 2PN, England.

First published in the United States in 1981 by
Facts On File, Inc.

Library of Congress Cataloging in Publication Data

World Energy ; The Facts and the Future
1. Power Resources I Euromonitor Publications Limited
TJ163.2.W69 333.79 81-658
ISBN 0-87196-564-X AACR2

Printed and Bound in Great Britain

"The world is running out of the fuels which built the technological world of today. Within a period which, seen in the perspective of the time taken to form our fossil fuels, would approximate to a few seconds, there will be very little oil, gas or coal left.

These few seconds, this eye-blink of geological time, are all that stand between us and the end of the world as we know it."

In this major new study, Don Hedley examines one of the most crucial issues facing the world today - energy - an issue which no country, business or individual can afford to ignore. With the aid of charts, graphs and statistical tables, all aspects of the current energy situation and the prospects to the year 2000 are extensively discussed and analysed; the structure of world energy supply; the fuels; the major users; usership of energy; the alternatives, and the future.

DON HEDLEY has worked for several years within energy industries and is at present employed by a major oil company. He lives in North London.

PREFACE

If you were asked what is the most serious problem facing mankind now, what would your answer be? Nuclear holocaust? Ecological catastrophe? The population explosion?

I believe that the most pressing problem is to maintain the world's energy supply. What has come to be known as the 'energy crisis' represents a threat to progress and to standards of living in rich and poor countries alike which is inevitable and imminent. Our fuels, the ones we use now, will run out and they will run out soon.

What is being done about it? What should be done about it?

'World Energy: The Facts and the Future' examines how energy is used in the world and how much energy is used; fuel resources - where they are, how long they will last, which countries have the fuel and which countries need it the most; the implications of the energy crisis for transport; the development of synthetics; the impact of conservation; the renewable energy sources and what progress is being made with them.

'World Energy: The Facts and the Future' forecasts how the world energy economy will have changed by the year 2000 and what is likely to happen beyond.

Energy and the way future supply is planned affects us internationally, nationally, commercially and personally. Our standard of living and our economic future are the stakes when the big energy decisions are made.

Don Hedley
London, January 1981

METHODOLOGY

There are many sources of energy usage data, and, whether for reasons of differing methods of interpretation, differing criteria or differing sources of raw data, there are frequent discrepancies when a number of authorities publish figures on the same topic.

This study attempts, where this occurs, to interpret consensus but where no consensus exists it has sometimes been necessary to arbitrate between reputable sources.

The approach used is to examine fuel use and energy prospects on a regional and major country basis and by analysis of each major fuel sector.

Predictions of future levels of consumption, and production of energy fuels in countries and regions are arrived at by taking account of - rate of consumption over the past 10 years; indigenous energy reserves; economic prospects - level of industrial development, population growth rate; government energy policy and general social and political 'atmosphere'; projections and predictions made by various bodies, institutions, government departments and individuals e.g. OECD, WAES, the World Energy Conference, the CIA and major energy companies such as Exxon and Shell.

A full list of sources appears on page 367

The practice of indicating future consumption of fuels by means of a range is not followed. This study recognises that there are a multitude of influences which can affect both the level and the nature of energy use but seeks to arrive at a best estimate based on the data available at the time of writing.

*   *   *

It should be noted that any views expressed are the author's own and do not represent the views or policies of any company, institution or organisation.

ENERGY UNITS

In this study energy usage is considered on a primary fuel input basis. The energy contained in the various fuels is expressed, for simplicity, and ease of comparison in the same unit - Million Tonnes of Oil Equivalent (MTOE). This represents the amount of energy contained in one tonne of oil.

100,000 tonnes of oil contains the energy to generate enough electricity for 50,000 dwellings for one year.

100,000 tonnes of oil could also be equated with the energy contained in the petrol used by about 50,000 passenger car drivers.

One million tonnes of oil is roughly equivalent, in calorific terms to:-

39 trillion (million million) BTu (British Thermal units)
395 million therms
10,000 teracalories
1.5 million tonnes of coal
3.0 million tonnes of lignite
5.3 million tonnes of peat
1,167 thousand million cubic metres of natural gas
41.2 thousand million cubic feet of natural gas
113 million cubic feet per day of natural gas for a year
12 thousand million Kilawatt hours (KWh) of electricity

Conversion of Oil Units

To convert tonnes of crude oil to barrels of crude oil multiply by 7.33 (one barrel consists of 42 gallons).

To convert tonnes of crude oil per year to barrels per day multiply by 0.0201.

To convert barrels of crude oil to tonnes multiply by 0.136.

To convert barrels per day of crude oil to tonnes per year multiply by 49.8.

CONTENTS

INTRODUCTION

KEY TO MAIN SYMBOLS USED

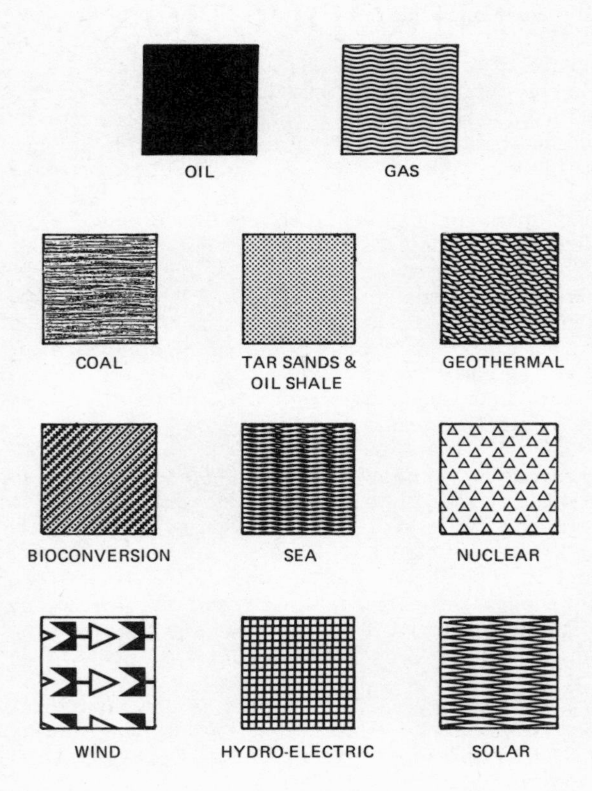

'World Energy, the Facts and the Future' is an attempt to simplify a complex subject - how the world is to supply itself with the energy to power its industry, heat its dwellings and generally fuel economic growth.

The essence of what has come to be called 'the energy crisis' is that demand for energy is exceeding economic supply. Energy demand growth is closely related to economic growth. Although the correlation between the two may be diminished by conservation and the reduction of the energy intensity of industry, the link cannot be broken. Industrial societies are geared to growth but even if a non growth equilibrium could be achieved, unless the population ceased growing, per capita energy consumption would have to fall, thus resulting in deteriorating standards of living.

However, world population will grow and, with it, the need for energy. Much of the energy demand growth will occur in the less developed countries whose present per capita energy consumption levels are many times lower than those in the industrialised countries. But whether or not these countries will be able to meet their demands, depending as many of them do almost exclusively on oil, depends on a number of factors.

The tremendous industrial growth of the nineteenth and twentieth centuries was made possible, in energy terms, by two transitions - the transition from wood to coal and the transition from coal to oil. The cheapness, abundance and flexibility of oil made it the main energy source and balancing fuel - the fuel to make good any new demand surge - whether for central heating in the home, petrol or diesel for transport or a new power station to augment electricity supply. The result was that the world became heavily dependent on oil and the dependence became a kind of technological trap.

The trap was sprung when the countries controlling most of the world's exportable oil reserves took over control of their oil from the companies who had discovered and sold it for them, and formed the Organisation of Petroleum Exporting Countries (OPEC).

The trap was lowered when a combination of circumstances, notable among which was the realisation that control of supply means the ability to fix price, gave rise to the leaps in oil prices (particularly following the Yom

Kippur war in 1973) which had a drastically damaging effect on the economies of all but the energy exporters. A sudden oil price rise causes a coronary in the world's energy arteries - immense sums of money are transferred from importer to exporter - huge balance of payments deficits are created at a stroke and the economic diseases of inflation and unemployment move from mild to chronic.

A further bar to the trap is that oil is a finite and diminishing resource. More is now being used than is being discovered. Thrown upon their own resources North America and Western Europe could meet their oil needs for only a few years. Further, the oil that is being discovered outside OPEC regions tends to be found in the harshest environments - thousands of feet below the sea or in Arctic regions - where production and development costs are high.

The next 20 years are crucial because during that time the foundations of another energy transition must be laid. This is the transition away from oil as the balancing fuel towards an energy economy where other fuels supply demand growth. The extent to which energy demand will increase is inextricably linked to the economic availability of primary fuels - coal, oil. natural gas and uranium.

The transition away from oil is underway, but the speed at which the alternatives are being developed is inadequate to secure the world against severe, oil provoked shocks to its economy. Furthermore, only in industrialised countries where the technology exists and the finance is available, can the development of alternatives be made a reality. The less developed countries (LDC'S) meanwhile are involved in the terrible 'catch 22' predicament of not having the money to invest in alternatives to stop bankrupting themselves by buying oil. The LDC's need financial help. The industrialised countries, while finance is still a major issue, have a further problem to find the political will to convince their electorates of the gravity of the energy situation and gain the necessary support to carry through unpopular measures such as vast expansions of the coal and nuclear sectors, two industries with particular environmental problems.

Energy is a commodity in increasingly short supply - unfortunately so is time - to build new mines, erect nuclear power stations, to sink new oil wells; all of these take a vast amount of time: needs 10, 20 and 50 years ahead must be anticipated. Unfortunately the Western World has been singularly unsuccessful in its planning ahead - the present perilous dependence on oil, a vulnerable resource of which the bulk of reserves lie in one of the globe's most politically volatile regions - seemed to come to pass unexpectedly. Furthermore, since the first blow fell in 1974/5 little has been positively achieved by the industrialised nations in the way of reducing that dependence and of taking the pressure off world oil supplies for the benefit of the LDC's.

Introduction

This study attempts to raise energy consciousness by presenting the facts of energy supply - what fuels are being used, by whom and at what rate. As well as analysing the supply prospects for each of the major fuels - the magnitude and availability of reserves, the main consuming nations and regions are examined as energy using units. Consideration is also given to the global economic trends related to crude oil prices, and a best estimate of world suppliable energy demand to the year 2000 is arrived at.

The transition taking place, albeit slowly, at this time from oil to a viable mix of economically available primary fuels is the precurser of another, hopefully more long lasting, transition to a millenium of renewable energy sources. The replacement to any significant extent of present energy producing technologies by inexhaustible sources, is many years and many billions of pounds of investment away while the financial and technological resources which can be devoted to their development are dependent on the success of present efforts at replacing oil.

The way to the hoped for renewable energy future has many potentially awesome pitfalls from energy famine to global conflict over depleting oil reserves. Energy is no peripheral government portfolio. The weightiness of the issue was clearly expressed in Chancellor Helmut Schmidt of West Germany's opening address to the World Energy Conference in Munich in September 1980 when he referred to energy as not only a 'central problem of human coexistence but also of the future of mankind as such'.

It is the contention of this study that this is no exaggeration.

Part One
THE WORLD

Chapter 1

The Energy, the Fuels, the Consumers and the Future

In 1979 the world used the equivalent of 6,960 million tonnes of oil, 3.2% more than in 1978, 40 times the 1900 level and four times as much as was used thirty years ago. The increase between 1978 and 1979, though lower in percentage increase terms than many previously recorded and lower in actual amount than the increase on several occasions, was nevertheless very considerable. At 217 MTOE it was the equivalent of adding the total consumption of the U.K. to the total.

The pattern of consumption over the past 10 years shows a steady increase up to January 1974, when the Yom Kippur war caused OPEC (the Organisation of Petroleum Exporting Countries) to double the price of oil and the world was catapulted into a recession. Since 1975, however, despite the continuing increase in energy prices, consumption has resumed its pattern of growth.

This study predicts a 75% increase in world energy consumption by the year 2000 - a stupendous increase greater than the world's total consumption in 1970.

Examining the fuels used to supply the world's energy from 1969 to 1979 shows how the world's dependence on oil grew. Until 1973 oil's share of world energy consumption increased to a point where it accounted for almost as much as all the other fuels combined. In 1969 oil had long superseded coal as the major energy source - its abundance, accessibility and adaptability brought almost limitless individual transport to the industrialised world. Just as once railways had shaped the patterns of urban development into ropes of dwellings clustered around railway lines so the motor car spawned the suburbs while ease of rural access softened the cities at their centres.

The rise in oil usage was checked in 1974 - for the first time volume consumption dropped - then dropped again in 1975. But despite the violent economic consequences of the change, caused by the doubling of crude oil prices by OPEC, the increase in volume consumption began again in 1976 and has been maintained ever since, even in the face of over twenty crude oil price rises.

The fact that this has happened - that the dangerous reliance on what has been proved a vulnerable resource has continued - is not however a testimony to the world's lack of foresight (though with the benefit of hindsight, energy planning over the past twenty years shows little appreciation on the part of governments of the economic realities of energy supply). The problem is that to change energy supply modes is a long and costly business. In microcosm this may be seen in the example of centrally heating a house in the U.K. Although anyone with oil fired central heating is aware that other fuels - gas or coal - are cheaper, the capital cost of the work necessary to substitute one system for another may render continued use of the more expensive fuel seems the most economical alternative. But in reality it is not, for oil prices will rise more quickly than alternative fuel prices and conversion costs also will increase.

Energy planning at Government level is more complicated than heating a house: other factors must be included in the equation. The first two might be availability and lead times: coal was the main energy source for many years. Now energy planners are predicting that it will be again. Ample coal reserves exist, as does the technology to mine the coal, but lead times - the time from the decision to produce to actual production - are long. A new coal mine can take between five and seven years before it becomes productive - old mines are worked out.

Other factors might be environmental - powerful anti-nuclear lobbies have already frustrated nuclear power programmes aimed at reducing use of oil in power stations.

Transportation of such fuels as coal and liquefied natural gas is in its infancy - neither the ships nor the port facilities exist.

Finally there is cost. In the U.K. a single nuclear reactor costs over £1 billion while the cost over the next 20 years of developing the alternative fuels to reduce the world's oil dependence, is estimated by the Dresdner Bank at $10 trillion (million million).

These are all mitigating factors - but reactions have been slow. Coal consumption since 1975 has increased less in volume than oil consumption though at a faster rate.

Since 1969, gas consumption has increased at a faster rate than that of either of the other two leading primary fuels and is the only one of the three which in 1979 was accounting for its largest ever proportion of world energy demand. Gas has the flexibility and ease of transportation of oil when used domestically. Its transportation by tanker in liquefied form, is less developed.

Furthermore gas reserves are far smaller than coal reserves* and the use of natural gas, should, like that of oil, be tempered by the thought of future scarcity and not squandered on uses where other primary fuels may ultimately be substituted.

Nuclear power, with uranium as the fuel, was hailed as the energy panacea when the first nuclear power station began electricity production at Calder Hall (U.K.) in the 1950's. Newspaper stories feted the ultimate substitution of ploughshare for sword. The awesome power that blighted the cities of Hiroshima and Nagasaki had now been harnessed for good. Scientists who would now blush at the memory spoke of free electricity. The reality of the exploitation of nuclear fission in a thermal reactor has proved somewhat different - in 1979 it accounted for 2.2% of world energy consumption, one third that of hydro electric (water) power - the only one of the so called 'renewable' sources of energy at present contributing meaningfully to world supply. Nuclear energy has suffered many setbacks and is at present in a fairly depressed state in every country but France and Japan. However, these setbacks will hopefully prove mere teething troubles because few attempts to plot a world energy future deny nuclear power a leading and integral role in the transition away from oil as the 'core' fuel in the powering of the world's economy.

By the year 2000 world energy consumption will be 75% above the 1979 level. The mix of primary fuels will have changed substantially. Drawing on its huge reserves and with the correct steps taken by governments and private industry to expand and rejuvenate the industry, coal should be the world's main energy source in the year 2000, accounting for 37% of demand. The end of the century will see significant international trade in coal with huge coal tankers plying routes between the United States and Europe, Australia and Japan.

Oil will make up a much smaller percentage of world consumption in 2000 than it did in 1979, at 33% as opposed to 45% but despite this, consumption will still increase by 29%.

That such an increase will occur should not be taken as anticipating the failure of governments to control an economically suicidal pattern of energy use. Rather, an increase in oil use held down to this extent will come only as a result of titanic efforts both to conserve energy, to make economic activity less energy intensive and to make transitions to oil alternatives. The inexorable

* Unlike oil reserves, world gas reserves are increasing. This could well continue until the end of the century.

increase in oil consumption to the year 2000 will not be subject to the control of today's main consumers, the industrialised countries. The bulk of the growth will occur in the less developed nations. It is unfortunate that the measures which can cut down oil use - measures such as conservation and the development of alternatives, are extremely expensive. The less developed countries are already significantly more oil dependent than the more developed ones. Unless prompt action is taken to make funds available, the energy crisis will cause the world's present economic inequalities to be exacerbated, with more and more of poorer LDC's exportable goods being used to buy less and less oil at higher and higher prices.

Not all the less developed countries are caught in the 'oil trap'. The oil exporters, notably the OPEC countries, are increasing their energy consumption at rates reminiscent of the industrial revolution in the West. Most of these increases are accounted for by oil and in the future the oil exporting LDC's will account for a significantly higher proportion of world consumption. Unfortunately however, increases in OPEC domestic consumption means less oil exported, more price pressure and more competition in the world market for a diminishing resource.

To 2000 therefore, there is no chance of actually reducing world oil consumption but keeping growth over the 1979 level below 30% will be a major achievement.

Natural gas consumption will increase by some 60% between 1979 and 2000, an increase which means gas's present share of world energy supply will fall from 19% to 17%. By the year 2000, the supply of oil, in some countries, of natural gas also, will be significantly aided by synthetic fuels. Synthetic fuels are liquids and gases which can be substituted for natural gas and conventional oil products in most applications. These can be obtained from coal, tar sands, and oil shale as well as from agricultural products.

Apart from coal, the only fuel increasing its share of world consumption will be nuclear power with a sixfold increase to account for 7% in 2000. Although nuclear power is not a diminishing resource in the way of oil or gas, predictions of the level of its future use certainly are. The level of nuclear reactor orders only one year ago might have justified a far higher level of installed capacity in the year 2000 than the one estimated in this study. Since then, a large number of cancellations have changed the picture somewhat but improving technology, rising energy prices and perhaps the occasional threat of energy shortage, should improve the public acceptability of nuclear power stations by a sufficient degree, to achieve the level of usage predicted here.

Public acceptibility will be an omnipresent influence operating to varying degrees on all changes in the nature of energy supply technology and modes. Nuclear power and liquefied natural gas raise questions of safety. Coal mines and the drilling and transportation of oil are major environmental issues, and in time the 'clean' renewable sources of energy such as wind, wave, solar, geothermal, biomass and fusion will have their own 'environmental' problems. Though at present, suffused in the public mind in a rosy glow of nostalgia for the gentle creak of wooden windmills and the general wish to adopt more simple, less technological life styles, the reality of machines designed efficiently to exploit natural forces and deployed in significant numbers will probably result in just as many 'anti-lobbies' as have the exploitation of the fuels that supply us at present.

The figures in Chart 1 for hydro power in the year 2000 includes the contribution of the new renewable sources. Most of the 65% increase from 1979-2000 will, however, be accounted for by hydro-electric schemes the main proportion of which will be in South America where the terrain offers the best potential for development.

In 1969, one single country accounted for one third of all the world's energy consumption - that country was the United States. To give this singular fact perspective a few comparisons might help. United States energy consumption then was roughly equal to that of USSR, China, Japan, West Germany and the U.K. combined and it was three times all the energy consumed in Latin America, SE Asia plus Africa, plus Australasia, plus S. Asia.

In 1979 the picture has changed but not a great deal. The United States is still the main consumer but its proportion of world energy consumption has dropped from one third to just over a quarter. The second largest consumer, the USSR, has increased its share of world energy consumption from 14.5% in 1969 to 16.5% in 1979. China, the country with the highest energy use growth rate amongst the main consumers, comes third. Of the top ten consumers, the U.K. has the lowest growth rate over the past ten years, in keeping with the low level of growth in its economy.

The Communist World has increased its energy consumption by more than the Free World over the past 10 years - from 26.1% of world energy use in 1969 to 31.1% in 1979.

Energy use in the developing world has increased at a much faster rate than in the industrialised world. But this does not mean that the LDC's are catching up - as a group they are falling further behind. The region which has grown the fastest since 1969 is SE Asia - where the pace of industrialisation has been particularly fierce.

Chart 1 **CONSUMPTION OF THE WORLD'S FUELS**

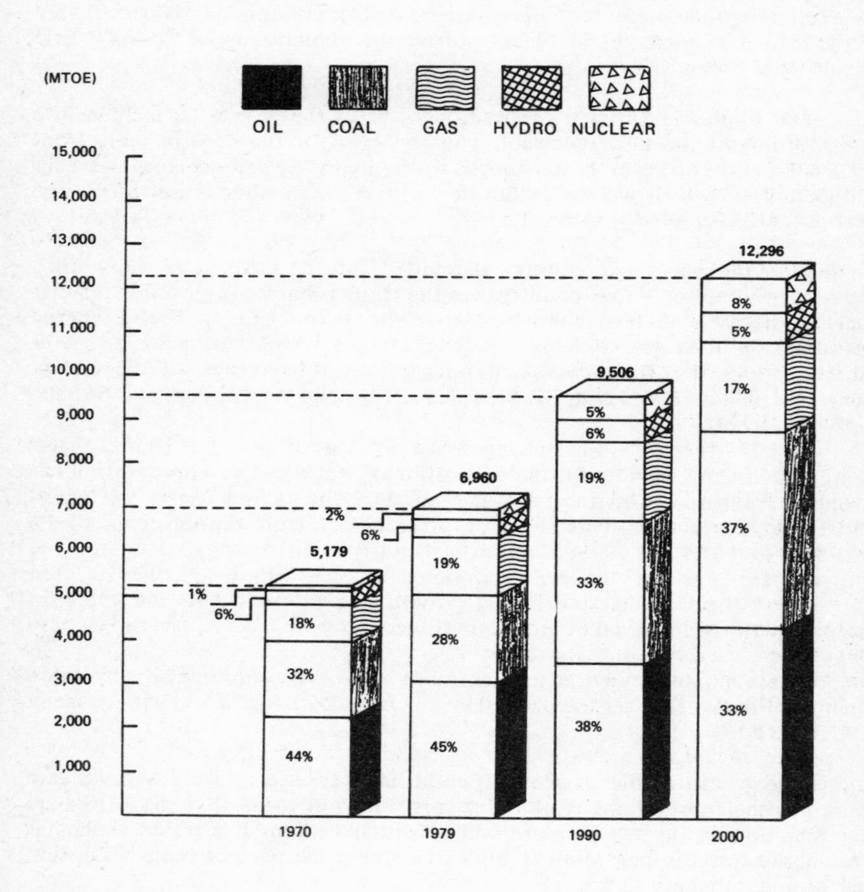

The key to the change in energy consumption which will take place to the year 2000 on a country and regional basis is that everyone will use more but that growth will be faster in the less developed countries. The fastest growth of all will come in the less developed energy exporting countries where fuel resources, notably oil, are in abundance. Thus, easily the quickest growth in energy use will be recorded in the Middle East - in the OPEC countries. Africa will be the region recording the second greatest growth over the period. Africa's growth will be large as a result of two factors, one - the level of per capita consumption in 1979 is particularly low in most African countries, two - what growth there is will come from the countries with their own energy resources e.g. Algeria, Libya, Nigeria.

The other LDC groupings China, SE Asia, South Asia and Latin America will also record high growth rates but these will do little to equalise the present disparities in both energy use and per capita productivity between the industrialised world and the developing world. Also, much of the decline in the rate of energy use in the industrialised world will come as a result of conservation and the general movement to less energy intensive modes of economic activity.

A further factor is that, in many less developed countries, predicted increased energy usage reflects no more than population growth while in some of the worst cases, population growth will outstrip economic growth leading to an actual fall in per capita productivity, per capita energy consumption and standard of living.

Looking to the industrialised regions and nations, Europe, the USA, Japan and the USSR will all experience slower growth in energy usage to the year 2000 than during the 1969-79 period. Indeed, all but Japan are expected to register lower growth also than during the 1974-79 period.

The USA, Japan and Western Europe were the hardest hit of the industrialised world when oil prices rocketed upwards in 1974. Japan, the most dependent of all on imported crude oil suffered the largest fall in energy usage, growth falling from an average yearly rate of 4.2% over the 1969-79 period to one of 1.5% over the 1974-1979 period.

The fact that the world energy depression will continue among the industrialised nations, with generally lower growth rates, should not be taken as indicating a continuing low level of economic growth, as this study expects the link between energy consumption growth and economic growth to become less absolute with conservation and changes of emphasis gradually altering established ratios.

Chart 2 THE WORLD'S MAJOR CONSUMERS

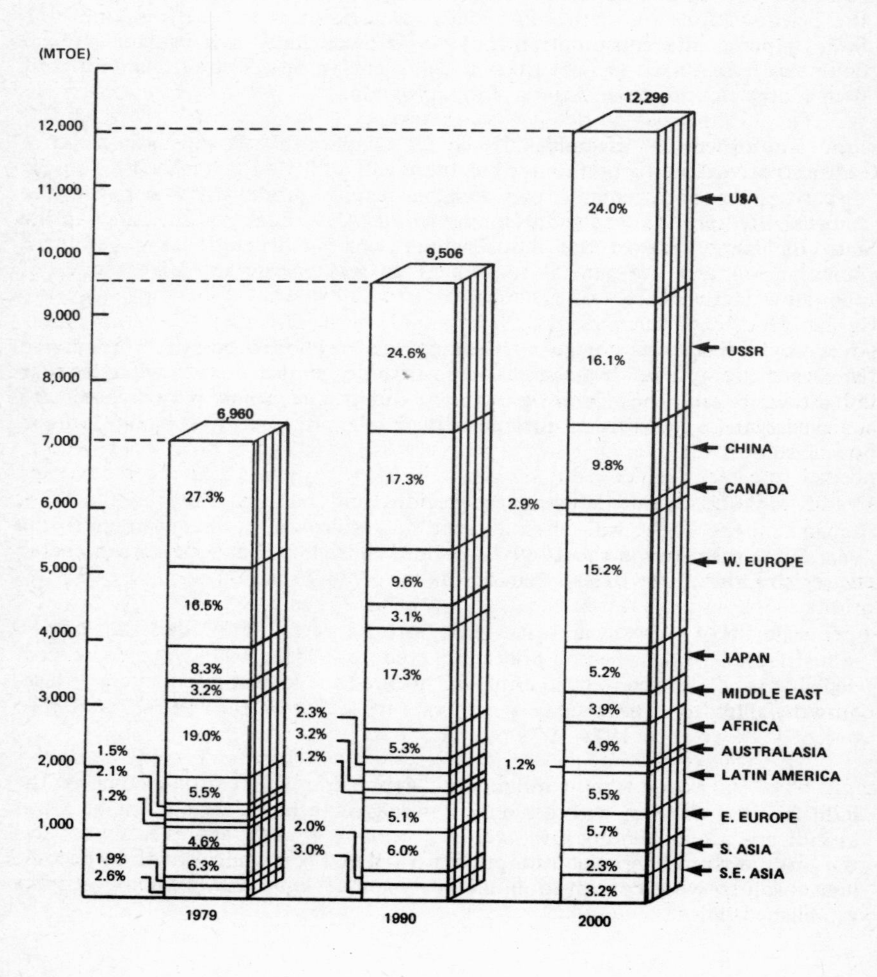

That oil is the life blood of most of the world's economic arteries is vividly demonstrated by analysing the proportions of primary fuels used in selected countries.

Of the twenty main energy consumers, in only China was oil not the main fuel in 1979 and for only eight of the twenty did oil not account for more than 50% of consumption.

The countries most dangerously dependent on oil were Japan (70%), Italy (68%), Denmark (79%), Sweden (64%), France (60%) and Spain (61%).

The Communist world was much less dependent on oil than the free world and as will be discussed later in the study, richer in reserves of fossil fuel alternatives to oil.

Of the selected countries examined, only the United States, France, Switzerland and Turkey reduced oil use between 1978 and 1979.

However, a number of the countries examined - USA, China, Japan, West Germany, U.K., France, Spain, Switzerland, Austria, Turkey, Denmark and Greece reduced the proportion of their energy demand met by oil. This shows the beginning of the big switch over from oil to alternative fuels in the industrialised countries. However, reductions in oil use of the size so far achieved tend to be fairly easily achieved - for example by lowering oil burn in power stations or changing work practices. The trouble is that the measure needed to effect further reductions and the substitution of other fuels become steadily more costly and difficult to implement.

Some countries are particularly well placed to meet the worsening world energy situation. Top of the list is Norway - where 70% of all power used is produced by hydro electric methods. Only 29% is met by oil and Norway is a net exporter of oil and gas.

The U.K. is also well placed, with indigenous oil, gas and coal and a comparatively low proportion of energy needs met by oil at 43% in 1979.

SE Asia is the most vulnerable region as far as oil dependence is concerned - oil consumption growing to account for 64% of total energy use in 1979.

The degree to which countries are able to change their mix of fuels in the years ahead to take account of the realities of fuel supply and economics, will bear significantly on economic performance.

The World

The USA, the main consumer, should succeed in lowering oil and gas demand to 47% of the country's total needs. A proportion of this demand should be met by indigenously produced synthetics, thus helping to lower dependence on imported fuels though these will still be necessary as conventional oil production falls. the two growth fuels which will fill the gap left by oil are coal and nuclear power.

Coal will be the main growth fuel in the USSR, which like the USA has huge reserves. Oil demand in the USA will be reduced to account for 31% of total energy demand in 2000 as opposed to 45% in 1979. The exploitation of the USSR's huge gas reserves will raise consumption in line with total energy demand and increase exports significantly.

China is likely to become more coal dominated during the next 20 years with the surplus oil produced from new offshore oilfields exported, bringing in much needed dollars to pay for industrialisation.

Western Europe and Japan should both achieve a less oil based mix by means of expansion of the nuclear industries and growing coal and gas imports. Japan with its paucity of indigenous resources will be a particularly heavy importer.

The Middle East, lacking the economic incentives of most other countries to reduce oil dependence in the short and medium term, will remain almost exclusively fuelled by oil with some gas. This commitment to oil, coupled with the growth in energy use, its cheapness and abundance make possible, adds to the seriousness of the world oil outlook. Although production in the OPEC nations will rise above present levels, domestic consumption will mean that less is available for export, putting pressure on world supply.

Latin America and South East Asia, despite some reduction in oil dependence during the next 20 years, will still be heavily committed to it in the year 2000 for about half their energy needs. Hydro electric power in South America, some nuclear power development in both regions and continued heavy coal use in SE Asia helps to reduce oil dependence. In South America, notably in Brazil and Venezuela, a proportion of oil demand will be met by synthetic fuels from biomass, heavy oil and oil sands.

Africa will remain a coal and oil continent, very much divided into haves and have nots with the haves expanding quickly as energy users and the have nots caught in an energy trap with more and more home produced goods being needed to meet oil import bills and less money consequently available to develop alternatives.

Chapter 2

Energy and Economic Growth

Although it tends not to be either simple or constant over a large time period, the relationship between energy demand and economic activity is a very important one. Increases in the output of similar goods using similar processes, will be accompanied by similar increases in energy consumption.

Also, evidence suggests that in most countries, growth in levels of personal income tends, subject to saturation effects, to be closely connected with higher consumption of energy.

The simplest expression of the relationship of energy consumption and economic growth is found in the concepts of energy coefficient,the percentage growth rate in consumption compared with the percentage rate of growth of GDP, or the energy ratio which is the ratio of energy consumption to GDP (Gross Domestic Product).

Neither of these concepts is a forecasting tool, rather attention to them following forecasting, indicates the net effect of such factors as change in energy mix, conservation and changing industrial emphasis.

The predicted level of growth in world energy demand from 1979-1990 is 2.8% while from 1990-2000 it slackens very slightly to 2.6%.

This basically reflects the predicted level of economic activity measured in growth in the world gross national product (GNP) and takes account of world population growth, much of which will take place in the less developed countries (LDC's). Between 1969 and 1979 world energy usage increased by the average yearly rate of 3.6%. The rate fell between 1974 and 1979 as a result of the 1974/5 oil supply disruption and subsequent price rise.

The level of world economic growth between 1979 and 2000 is expected to average just over 3% per annum which is marginally above the rate between 1973 and 1979. This low rate of growth contrasts with the rate of over 5% experienced between 1965 and 1973; it is only marginally above the 1973-1979 recession period rate because population growth will slow and because of sluggish industrial growth.

Chart 3 **WORLD ECONOMIC GROWTH OUTLOOK**

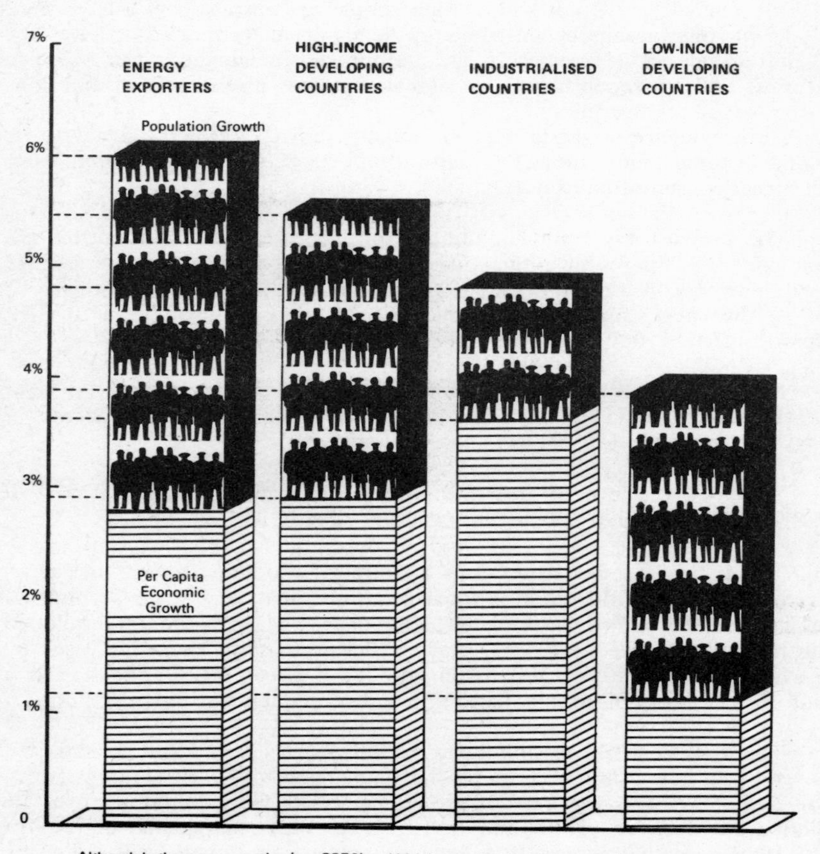

Although both energy exporting (e.g. OPEC) and high-income developing countries (e.g. Brazil, S. Korea) will grow, in GNP terms, faster than the industrialised nations, in real standard of living terms the main producers will do best and the gap between rich and poor nations will widen.

The fact that the predicted level of growth in energy usage between 1979-2000 is lower than that between 1973 and 1979 reflects less energy intensive economic activity and the growing importance of conservation measures and more efficient use of energy. This reduction in energy consumption per unit of GNP will be in the industrialised countries who have been lowering the energy intensity of their industry since the early 1970's. Reductions in energy intensity can be achieved by conservation - making do with less energy or by being more efficient - doing the same job with less energy.

Reductions in energy intensity are also occurring in the industrialised countries as a result of changes in industrial emphasis towards service industry and away from heavy industry. On the other hand, the emphasis on heavy industry which accompanies industrialisation will cause the LDC's inevitably to increase the energy intensity of their operations.

In the industrialised countries, conservation measures, such as adjustments to the thermostat settings of buildings, the fitting of insulation, improvements in the internal combustion engine to give improved mileage performance, will continue the trend towards less energy intensity. Thus the slowing in the growth of per capita energy usage in the industrialised countries as compared with the energy exporters and the other less developed countries as we look ahead to the year 2000 will not mean a reduction in standard of living of the industrialised countries. For example, the growth in energy usage to the year 1990 of the USA should be seen in the context of a reduction of around 20% in the amount of energy consumed per unit of GDP as compared with 1969. This figure could rise to around 30% by the year 2000.

West Europe, not so heavy a per capita user of energy as the USA and Canada should achieve reductions in the amount of energy used per unit of GDP of about 20% by 2000.

At the Venice Economic Summit the 'breaking of the link between economic growth and the consumption of oil' was established as one of the main priorities for the world's leading nations. To achieve this, it was stated, 'maximum reliance should be placed on the price mechanism and domestic prices for oil should take into account representative world prices'. It was added that 'market forces should be supplemented where appropriate by effective fiscal incentives and administrative measures'.

Some recent thinking on the subject suggests that providing that high energy prices are maintained and conservation encouraged, the oil use/economic growth connection may be easier to sever than has previously been supposed for the reason that a great deal of energy use is only indirectly related to economic growth. One example of this kind of energy use is

Chart 4 **WORLD RESERVES 1979**

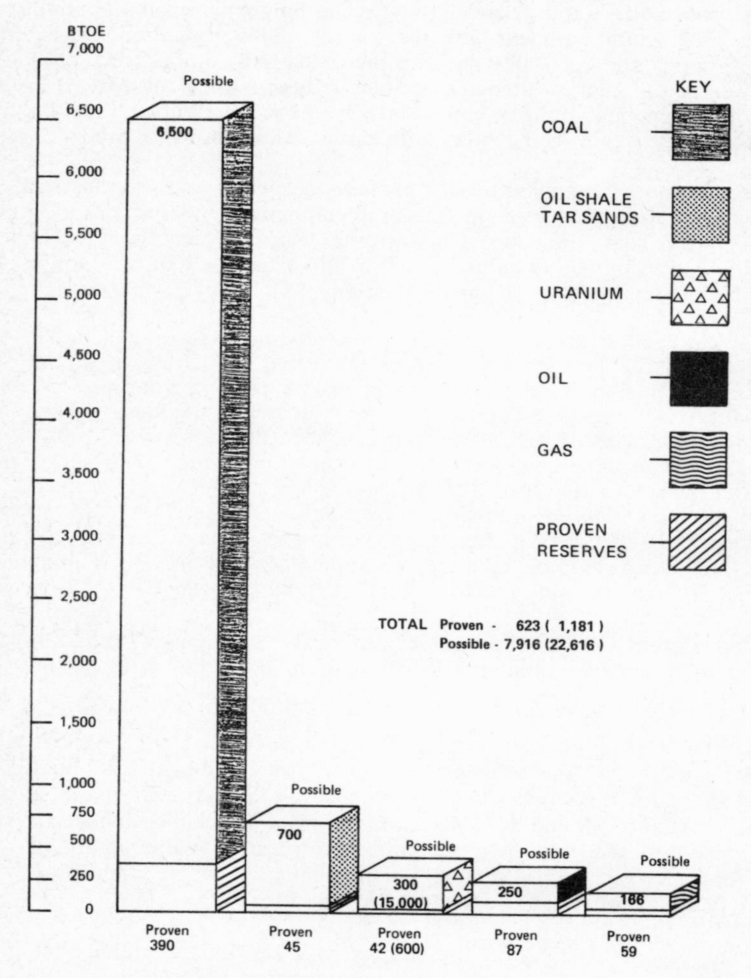

(Figures in brackets indicate reserves with use of fast breeder reactor)

transport and clearly there is abundant scope for improved efficiency which could reduce energy use without damaging economic growth prospects.

However in this study it has been assumed that for every ½% variance in the average economic growth rate during the years 1979-2000 between 500 and 600 million tonnes of oil equivalent per year might be added to or subtracted from demand estimates.

Reserves - How Much Fuel Do We Have Left?

Firstly, the world really is running out of the fuels which built the highly technological world of today. Within a period which, seen in the perspective of the time taken to form our fossil fuels, would approximate to a few seconds, there will be very little oil, natural gas or coal left.

But this few seconds, this eye blink of geological time is all that stands between us and the end of the world as we know it. It is a breathing space.

The most disquieting projection in the whole of this study is as follows:

Taking all the world's proven fossil fuel reserves - oil, coal, natural gas, oil shale and tar sands and adding on the energy which could be derived from the world's proven reserves of uranium we find that the world has the equivalent of 623 billion tonnes of oil to supply its energy needs.

In 1979 about 6.96 billion tonnes of oil equivalent were consumed. The projected level of increase in world energy demand to the year 2000 is 2.7% annum.

Dividing proven reserves by projected consumption we find that by the end of the year 2000 32% of our reserves would be gone, in 31 years 80% would be gone and by the year 2023 the world's energy tank would be empty. Thus a child born in 1979 might witness, at about the age of 40, the collapse of all the political and economic institutions we know today.

This will not be allowed to happen. Possible fuel reserves, particularly those of coal are far higher than proven ones while the energy derived from proven uranium reserves could be increased thirty fold by the widespread adoption of the fast breeder reactor. However, the fact that our present proven energy resources can give the world no more than another 40 years of civilisation is a grim reminder that we are not doing enough to free ourselves from our perilous dependence on fossil fuels. The truth is that lack of foresight and chronic complacency on the part of Governments has involved the world in a deadly game of ecological brinkmanship; our preparations for winter are

Chart 5 **WORLD RESERVES : PRODUCTION V RESERVES**

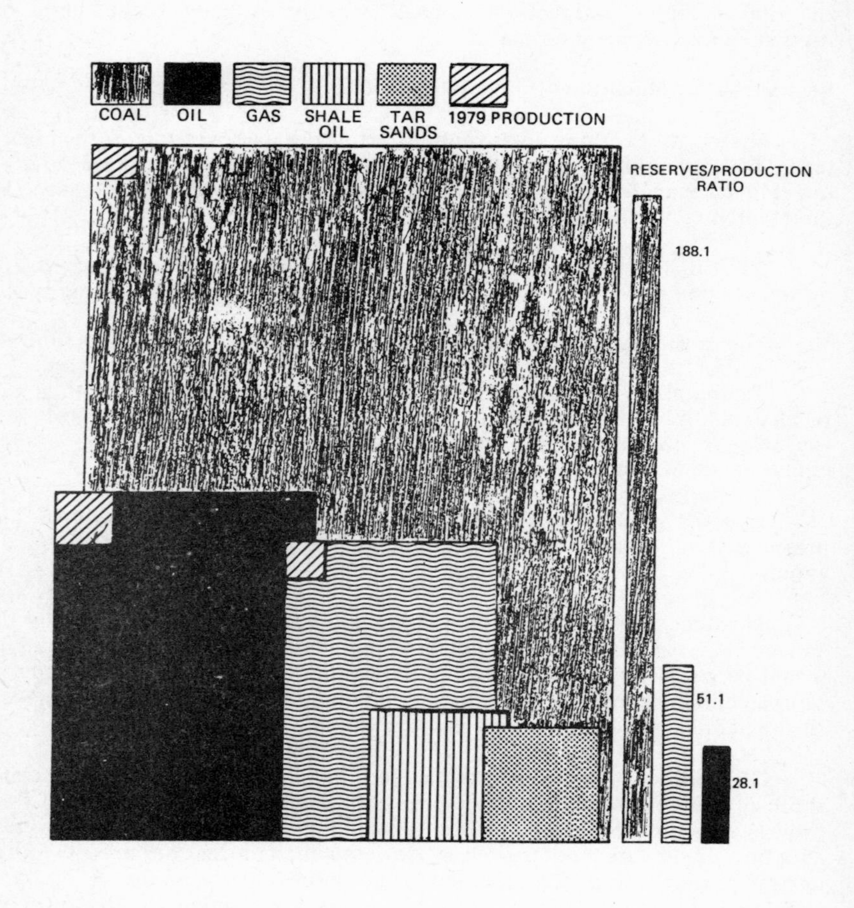

Source : BP

Chart 6

DEPLETION PROSPECTS*

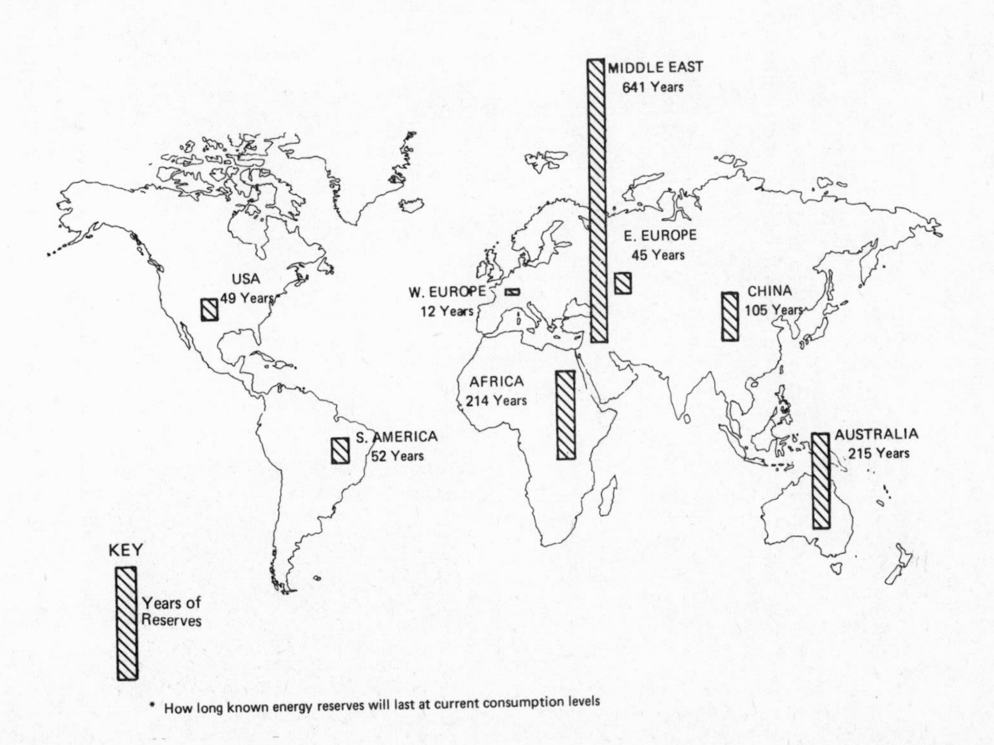

* How long known energy reserves will last at current consumption levels

being made at the end of autumn and if we are not very careful we are likely to freeze to death.

The extent to which our present pattern of energy use is out of step with the realities of supply may be seen by a comparison of the depletion prospects of the fuels which supply our present needs.

Even if the present rate of consumption were not exceeded, proven oil reserves will be 80% consumed in 22 years. Gas reserves, which are about 30% smaller than oil reserves would be 80% used up in 36 years. The coal situation is more encouraging, proven reserves would last for 158 years at present rates of consumption but the transition from oil to coal as the world's main fuel has scarcely begun and, as our projections show, there is one commodity in even shorter supply than oil - that commodity is time.

Constantly in evidence in any analysis of the energy situation is the fact that the geographical disparity between local use and local resources of energy is at the root of many of the day's economic and political problems.

Western Europe is the region most vulnerable in its combination of large consumption and small reserves. Thrown upon is own resources Western Europe would only have the energy reserves to keep going for 12 years.

The country with the worst reserves/consumption ratio in the world is Japan, with huge consumption and virtually no indigenous fossil fuel reserves. While, at the other end of the scale, the Middle East has the resources to support itself for more than six centuries at present rates of consumption without discovering one more oil or gas field.

Note: A number of different sources have been used for the Energy Reserves Section of this study and those parts of the study dealing with particular fuels. This is to demonstrate that there is wide measure of agreement on this vital subject.

Part Two
THE FUELS

Chart 7 COAL RESERVES 1979

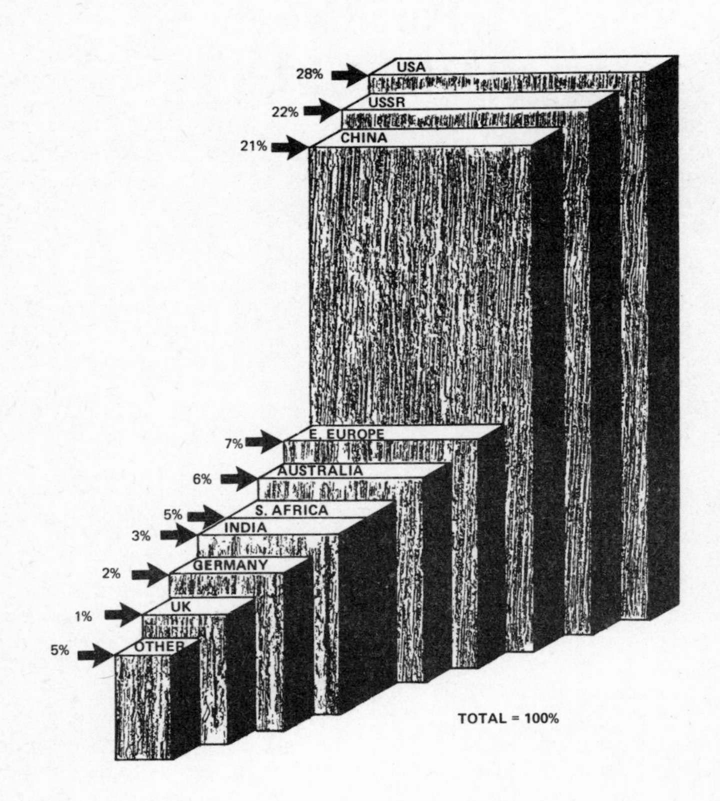

Chapter 1

Coal

According to the highly influencial book by the 'World Coal Study'* coal is to be the 'bridge to the future'. This means that coal is seen as the means by which the energy gap left by declining oil supplies may be filled, that coal can take over oil's role as the balancing fuel, ready to turn its BTU's and feedstock molecules to any use. The future to which coal is offered as the bridge is one where renewable sources will take over. The basic arguments are that oil is running out, so is natural gas; nuclear power has acceptance problems and the renewables are many years and many billions of dollars of investment in research and development away.

Building the bridge will not be easy - there are big problems. Not the least of them is the resurrection problem - the fact is that the world coal industry, except in a few locations, was allowed to run down with the rise of oil. Oil had all the advantages - it was cheaper to mine, there were fewer risks involved on getting it out of the ground, it was easier to transport, more adaptable to different uses and its energy content was higher.

Another set of problems that coal has if it is really to become king again are environmental. The basic point is, as the axiom has it, that coal is a great energy source except that no one wants you to mine it, move it or burn it.

Coal mining can be very disruptive to the landscape. Deep mines are tremendously expensive to sink and run, the future is likely to veer more towards 'factory mining' with huge open coal sites, but these, despite undertakings to return land to its previous state, create long term environmental disruption.

A further problem is the money: a whole new infrastructure must be created almost from scratch if coal is to be, as oil has been, a commodity traded on a world scale. Power stations to make the coal into usable energy must be built and this will be one of the most significant items of capital expenditure. For there seems very little likelihood of coal ever becoming again a major domestic fuel on the scale of former times at least in the industrialised world. The cost of developing the US coal industry to the extent

* 'Coal - Bridge to the Future' Ballinger Publishing Company, Cambridge, Mass.

that many projections deem necessary has been put at $116 billion* (in constant 1977 dollars). This would be for 700 new mines over the next 20 years, the construction of ten 1000 mile long slurry pipelines and 8,200 new railway trains, 300 barges and 16,000 lorries.

The Dresdner Bank, in a paper to the 11th World Energy Conference estimated that out of the $10 trillion dollars the Western World would have to spend on new energy equipment over the next twenty years $805 billion would be spent on the coal industry.

All these factors might make one sceptical of the world coal industry's grandiose predictions of its future importance but for one thing: there would seem to be no practicable alternative.

That coal is dirty, expensive and inconvenient is the bad news but this is outweighed by the good news - that there is a lot of it about. Estimated total resources (not the same as proven reserves) of coal are 26 times as large as those of oil and the technology to mine it is tried and tested. The technology to extract energy from coal is well developed and though the cost will be very high, coal can duplicate oil in most of its uses - even as a transport fuel. The manufacture of synthetic fuels from coal by liquefaction technique, will make a growing contribution to world oil demand during the latter years of the 20th century.

So coal's destiny as the world's major fuel is at hand - the bridge to the future is essential though it may be by virtue of its qualities of inconvenience and expense, a bridge of sighs.

But coal will never achieve oil's preeminence. The traffic on the bridge to the future is too heavy for that and other fuels must shoulder some of the load.

The world's coal reserves, though generally not to be found near the main oil reserves are similarly concentrated in a fairly small group of countries.

The USA has the largest economically recoverable reserves accounting for 28% of the world total, followed by the USSR and China. Between them these three countries, 'the Saudi Arabia, Iran and Iraq of coal' have 70%. Australia, South Africa and Poland are also major coal provinces and as coal use increases and a world coal trade grows up, will become major exporters.

The situation is slightly different when one examines the world's

* World Energy Conference

geological coal resources. 45% of this falls within the Soviet Union, but much of this, like Soviet gas reserves, is in the far North East of Russia where climate and distances cause significant problems.

The world's coal resources are much less equitably divided than world oil reserves: almost 90% being within the boundaries of only four countries - USSR, United States, the Peoples' Republic of China and Australia, while 10 countries account for 98%.

It should also be remembered that there may possibly be a great deal more coal in the world than has, so far, been discovered. Reserves increased by about 40% in the five year period to 1977 as a result of the renewed interest in exploration stimulated by the crude oil price rises of 1974.

A further important factor is that although it is improbable that major new coal provinces will now be discovered in the countries with histories of coal exploitation i.e. those for whom coal provided the energy for industrial revolution, the potential does exist for discoveries in the developing countries.

The story of coal consumption between 1969 and 1979 highlights the distinction between the industrially sophisticated and the less developed. In Western Europe and North America the percentage of world coal use accounted for fell, as did the actual amount used, while the rest of the world, to varying degrees, increased coal use.

During that period the Communist world overtook the non Communist world as coal consumers - the Communist bloc, in 1979, accounting for more than half of world coal consumption.

This trend will be reversed during the next 20 years with the non communist world increasing its coal consumption faster than the Communist countries. The World Coal Study suggested that a threefold increase in coal use was necessary by the end of the century.

This study predicts a 130% increase in world coal consumption with a 153% increase registered by the non Communist world.

Large percent increases will be registered in Africa, Latin America and South Asia (mainly India) but the bulk of the vast increase in consumption which is to make coal the world's premium fuel must come from the United States.

The USA's coal consumption increased by only 17% in the 10 years between 1969 and 1979. To accelerate this rate of growth to achieve a 200%

Chart 8 WORLD COAL CONSUMPTION 1979-2000

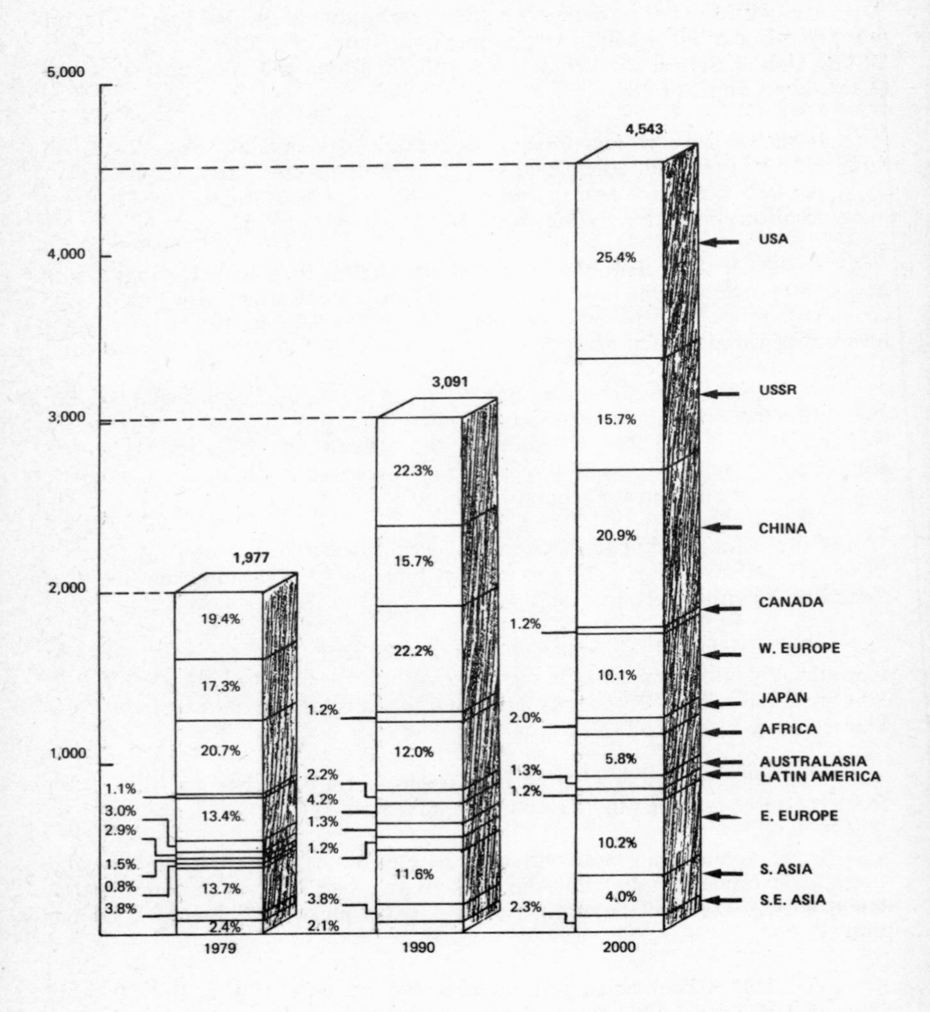

increase by 2000 will involve a vastly expensive change of energy infrastructure.

Many of the European countries - West Germany, U.K., France, Belg-Lux, Netherlands, Switzerland, Austria, Norway - decreased coal consumption between 1969 and 1979. The years to 2000 should see a reverse of this process with all consuming much more coal. The two main consumers, (those possessing significant reserves) West Geramny and U.K., will both register increases of just under 50% while Sweden and the Netherlands, both countries with supply problems caused by changes in energy policy will have most need of coal as a balancing fuel.

Japan, the country with the least indigenous resources compared to consumption of energy, will also need to increase coal consumption significantly. This increase will have to come from imported coal.

Just as it is today, the main use of coal in the year 2000 will be as a steam raiser in power stations.

However, there will also be a growth of coal use in industry, while a new and growing market for coal as a feedstock for synthetic oil and gas plants, a field under intense development at present, will grow in importance during the 1990's.

In Europe about two thirds of the increase in coal demand over the next two decades will have to be supplied by imports. The main increase in demand will be for steam coal, the proportion of metallurgical coal will decline due to the continued contraction of the European steel industry.

Industrial use of coal will increase as it replaces oil as a heating fuel and the overall percentage of coal, not used for electricity generation or metallurgically will also be swelled by use in liquefaction and gasifcation plants.

The fall in coal use in industry since 1950 should be reversed towards the end of the 1980's with coal use particularly increasing in cement, chemicals, petroleum refining, pulp and paper industries.

There will also be much use of coal in fluidised bed consumption the technique in which solids can be made to behave as fluids and which enjoys better heat transfer characteristics than present pulverised fuel furnaces.

The market for coal feedstocks for synthetic oil and gas will ultimately be a highly significant one. It has been estimated that there will be at least 20 such plants in the OECD alone by 2000.

The main area of coal use - electricity generation - will increase as a result of modest continuing growth in electricity demand and the substitution of coal (and nuclear) for oil.

Japan will experience the highest increase in coal generated electricity by the end of the century from 4% in 1977 to 16% in 2000. The level of coal generated electricity as a proportion of total electricity generation will also increase significantly in North America and Australia but will remain level in Western Europe due to nuclear power growth.

The nine largest coal producers, USA, USSR, China, Poland, West Germany, U.K., Australia, South Africa and India accounted for about 85% of production in 1979. This proportion is likely to change little over the next 20 years. Of these nine only India will have used over 20% of its recoverable coal reserves and of the world's present economically recoverable reserves only 15% or so will have been used.

But the problem of increasing coal production is not one of reserves but of inertia and investment. In Europe, coal production has declined steadily for 20 years - in 1979 it was 80% less than in 1959 and production levels at present are significantly lower than those before the energy crisis of 1974.

The EEC is particularly concerned that the European coal mining industry, despite the abundance of the resource, has contracted and a four point plan is now under consideration to increase investment in new mines; introduce a substitution programme for coal to replace oil in power stations and industry; adopt a policy fostering indigenous production and inter-community imports, increase the funds available for research and development of mining techniques, coal combustion, liquefaction and gasifaction.

If coal is in any significant sense to replace oil as the world's balancing fuel over the next twenty years, a world coal trade must be swiftly established.

The viability of coal as an international energy source depends on the ability of producing countries to transport and export large quantities.

The demand for imported coal will increase by just under 300 MTOE between 1977 and 2000 to about 413 MTOE.

The largest increases will come in the developing areas - Africa, Latin America, the East and Asia. These countries will account for about a quarter of all coal imports in 2000.

OECD Europe will increase its demand for imported coal almost fourfold with particularly heavy increases in imports to France, West Germany, Italy, the Netherlands and Sweden.

Japan will be the largest single coal importer at 88 MTOE, which is more than the U.K.'s total coal consumption in 1979.

Steam coal will make up the bulk of the imports at 56% in the year 2000 with Japan the main importer of both steam coal and metallurgical coal. Metallurgical coal imports will increase most to the developing countries where it will be used in the growing heavy industries.

Much depends on the USA, Australia and Canada if the world coal trade is to take off. Between them, these three should be responsible for over 65% of exports in the year 2000 (as against 50% in 1977). Other significant exporters will be South Africa, Poland and the USSR. India and Indonesia, both fairly large producers, will have little surplus for export.

The main direction of world coal trade flow will be from Australia, Canada, Poland, the USA and South Africa to Japan, Asia and Western Europe.

Some exports will come from developing countries - but these will take time to build up. The major oil companies have been in the coal business for some time and exploration for and development of new supplies is taking place all over the world. Exxon, the world's largest oil company, for example, has a coal project with Carbocol, the Colombian state coal company, to develop an estimated 1.6 billion tonnes of steam coal in the Guajiva Peninsula with production estimated to start in the mid to late 1980's.

The coal market will be the world's largest energy market by the year 2000 but the amount of political and financial commitment necessary to increase production and change a largely indigenously used and produced fuel, into a 'swing fuel' moving freely around the world to balance energy shortfalls should not be underestimated.

To achieve the level of trade required, perhaps 1000 ships of some 100,000 tonnes (dwt) and each costing about $40 million will be needed.

As well as this new ports, terminals and rail facilities will have to be built.

Many of the technologies for producing, moving and using coal are tried and trusted but there is no doubting the problem in martialling and committing the massive sums that will be necessary.

In all this the role of the multi-national oil companies is a vital one, particularly as regards the establishment of a viable world coal market. Long experience in the vertical integration of the oil industry gives them a clear advantage over governments.

The major oil companies' diversification into coal is well known: BP is the eighth largest private sector coal producer in the world and it has been estimated that by 1985, oil based companies could control about 40% of US coal output.

Oil companies have traditionally demonstrated an ability to get things done quickly - an ability which will be of prime importance if the coal industry is to be expanded in time to meet world demand and fulfil plans, such as the one drawn up by the Coal Industry Advisory Board of the IEA to triple the West's use of coal by 2000.

Long lead times for gaining planning consents for mines, establishing transport facilities and building power stations (the most expensive single link in a coal chain at about $1 billion per 1000 MW) have always hampered the development of the coal industry and will, no doubt, continue to do so, particularly in the US and Europe, but it is to be hoped that serious disruption of the industry's ambitious plans will not occur. The transformation of this abundant and familiar energy source into the world's main energy source by the year 2000 is a vital ingredient of the world's maintaining an acceptable level of economic growth over the next 50 years.

Chapter 2

Natural Gas

Natural gas is one of the growth fuels of the moment. Unlike oil more is being discovered every year than is being produced and at some point, probably before the end of the century, gas reserves could exceed oil reserves. But like coal, gas must become more of a world traded commodity than it is at present for its usefulness to be maximised.

Natural gas possesses many of the advantages of oil: once discovered it is comparatively easy to produce. As a fuel it is flexible in use and easy, though more expensive than oil, to transport by pipeline. It is also a hydrocarbon and thus, can be substituted for oil as a chemical feedstock in many applications.

Over the past ten years natural gas consumption has increased more than that of either oil or coal but it has been an increase based on indigenous use rather than trade. All the major energy consuming countries relying on gas for more than 20% of their energy needs - USA, USSR, Canada and the Netherlands, are major domestic producers.

Proven reserves of natural gas were, in 1979, 68% those of oil with 36 years consumption at present rates until 80% depletion.

World gas reserves constitute about 10% of the world proven primary fuel reserves, more than the potential energy content of the world's proven uranium reserves (if used in conventional nuclear power stations).

Natural gas is a growth fuel. Over the past 10 years international deliveries have almost quadrupled to 150 MTOE and, by virtue of existing contracts will increase by about 60% to 240 MTOE by 1985, but once the LNG (liquefied natural gas) trade begins to become a significant presence it could easily become the target of environmental groups and safety lobbies.

A transportation infrastructure is the first requisite of continued growth in gas consumption. LNG transportation by sea seems a less promising contender as the main medium of transportation and the future seems now to belong mainly to the international gas pipeline.

Chart 9 LOCATION OF NATURAL GAS RESERVES 1979

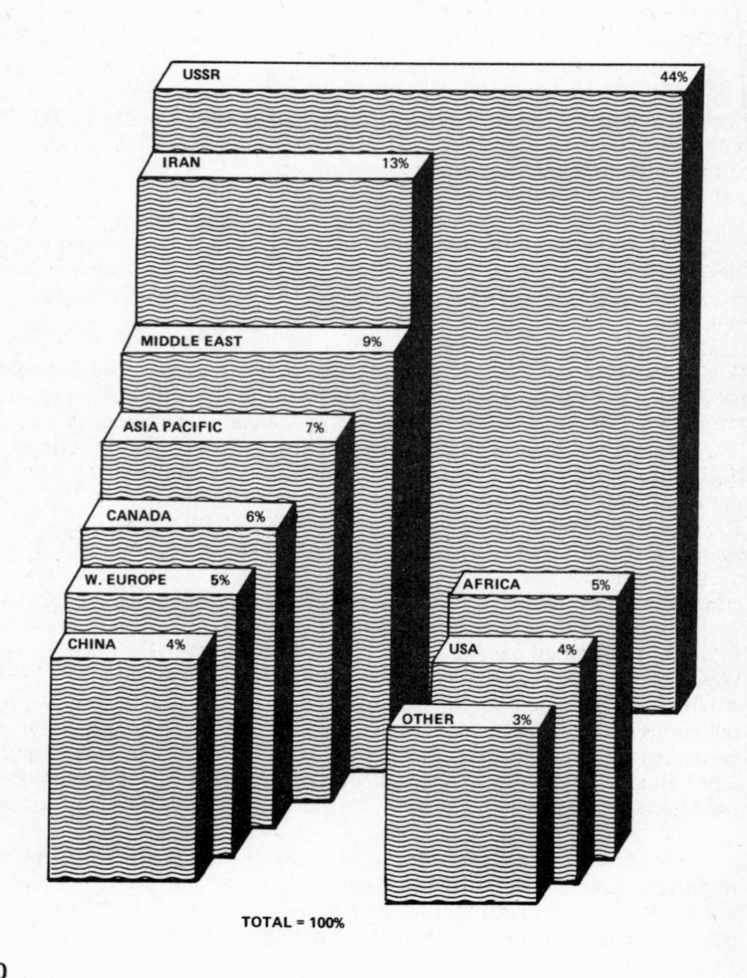

The main consumers of natural gas are the US and USSR, accounting in 1979 for 64% of world consumption, however USA consumption has fallen by 11% over the past 10 years while that of the USSR has risen by 112%. The fall in United States consumption is likely to continue though the decline will be gradual.

One of the main growth areas since 1969 has been Western Europe where the discoveries off the Netherlands and in the U.K. and Norwegian sectors of the North Sea have had a radical effect on consumption patterns.

The Netherlands is a particularly striking example of a mismanaged energy bonanza, though, of course only hindsight makes such judgement possible. In 1959 the vast Groningen gas field was discovered and the Netherlands became heavily reliant on gas - 85% of Dutch homes use gas heating and 80% of industry uses gas.

The Netherlands is also the world's largest gas exporter, supplying about 40% of the needs of Western Europe. The revelation of the 1973/74 energy crisis showed the Dutch that their gas was not, as they had viewed it, a way of making a few 'fast bucks' before other energy sources such as nuclear power rendered the gas fields suddenly obsolete, but a finite and valuable resource that they were squandering on low price exports which did not reflect either present value or future scarcity.

The main difference between world gas reserves and world oil reserves is that gas is growing and oil is diminishing. In other words more oil is being used than is being discovered - in 1975 proven oil reserves were 89.5 BT while in 1979 they wre 87.3 BT. Proven gas reserves, on the other hand have doubled since 1970 and are now 63.5 BTOE. Proved and undiscovered gas reserves are usually put at about 172 BTOE some 100 BTOE less than estimated cumulative oil reserves but this figure also is constantly being revised.

A conference in summer of 1980 on Conventional and Unconventional World Natural Gas Resources at the International Institute for Applied Systems Analysis in Vienna was told by M.S. Modelevsky that the total value of the resources economically recoverable under current and near future conditions come to about 250 tcm or 215 BTOE. It is therefore probable that gas reserves will soon exceed oil reserves and large gas consumption increases are the obvious corollary.

Modelevsky also predicted that world gas consumption over the next two decades would be 2.5 to 3.0 times greater than total production to 1980.

Even such a rate of consumption would not mean serious depletion and the limiting factors for the growth in use of gas are thus technical and economic factors rather than geological ones.

The USSR possesses more than 40% of both proved and possible reserves and will become the world's largest exporter of gas in the fairly near future depending on the progress of new pipelines.

Canada and the USA, between them account for about 10% of world reserves (proven and possible) while the North Sea discoveries, make up, along with the rest of West European gas reserves, about 5% of the world total.

Apart from the USSR the most significant deposits of natural gas are to be found in the Middle East which accounts for more than 20% of reserves.

The USSR will become the world's largest single exporter, quickly overtaking the Netherlands, partly as a result of the construction of an East-West pipeline network carrying gas from Western Siberia where the huge fields of Urengoiskoye, Yamburgskoye, Zapolyamoye and Medbeshye alone account for 9.2 billion tonnes oil equivalent (14% of world proven reserves).

However, the OPEC countries of Iran, Algeria, Saudi Arabia, Kuwait, Qatar and Libya will increase rapidly as exporters, particularly of LNG and by 1990 could be accounting for some 37% of the world gas trade.

Such a hold on world trade would give the OPEC gas exporters the control of gas movements which could lead to the same price controlling cartel which exists in the world oil market and the USSR would be unlikely to resist the call already being heard among the main exporters that gas prices should follow those of crude oil.

The United States is the main natural gas producer in the world accounting for 37.5% of the world total, followed by Russia. These two account for 65% of world production.

Production in the Netherlands the third largest producer, peaked in 1977, fell significantly in 1978 but registered a 6.1% increase in 1979. Canadian production has followed a similar pattern as has that in the U.K. whereas Indonesian and Chinese production are increasing steadily.

USSR's production is likely to be limited only by availability of the financial resources to develop its huge and often distant reserves. 1979 saw an increase in output mainly from Central Asia, Western Siberia and the Ukraine and a similar increase is forecast for 1980.

World production of natural gas will continue to increase with the rate of growth probably accelerating past the mid 1980's.

Production in the United States will inevitably continue its decline but this decline will not necessitate a paralled fall in consumption as imports from the North (Canada) and from the South (Mexico) will increase.

Latin American production will increase mainly as a result of greater output from Mexico, Venezuela and Argentina. Mexico will be the largest South American producer - Pemex, the country's government owned oil and gas company is planning an increase of 34.1% in output in 1980 to meet US plans for pipeline imports.

Middle Eastern output will, in the short term, remain at a fairly low level due to the war involving Iran, the main producer.

African output will increase quickly, mainly as a result of Algerian and Libyan production. These will be in the near term the most significant LNG exporters achieving over 1.72 MTOE (2 billion cubic metres) per year. Production in Nigeria will grow as a result of a new gas gathering system.

By the year 2000 the USSR will be easily the world's largest natural gas consumer.

The fastest growth in usage however will be in Africa, the Middle East and Japan. Africaand the Middle East's consumption will be concentrated in the producing countries themselves, helping to fuel industrial growth. Japan's will be made possible by the growing world LNG (LiquefiedNatural Gas) trade.

Growth will also be high in South East Asia and Australasia but in neither of these regions will consumption achieve significant levels. Other than Japan the main importers will be the Western European countries.

United States consumption has declined since 1973, in parallel with production. The completion of the Alaska pipeline to supply 48 states will slow the fall in consumption considerably.

In Western Europe, indigenous supplies will increase to the mid 1980's after which demand growth will be met by imports, particularly from USSR, Algeria and the Middle East.

In 1979 about 1,297 MTOE of natural gas was used as opposed to being reinjected or flared. Of this 1,153 MTOE (89%) was used in the country of production while 144 MTOE (11%) was traded, of which 22% was made up of

Chart 10 NATURAL GAS CONSUMPTION 1979-2000

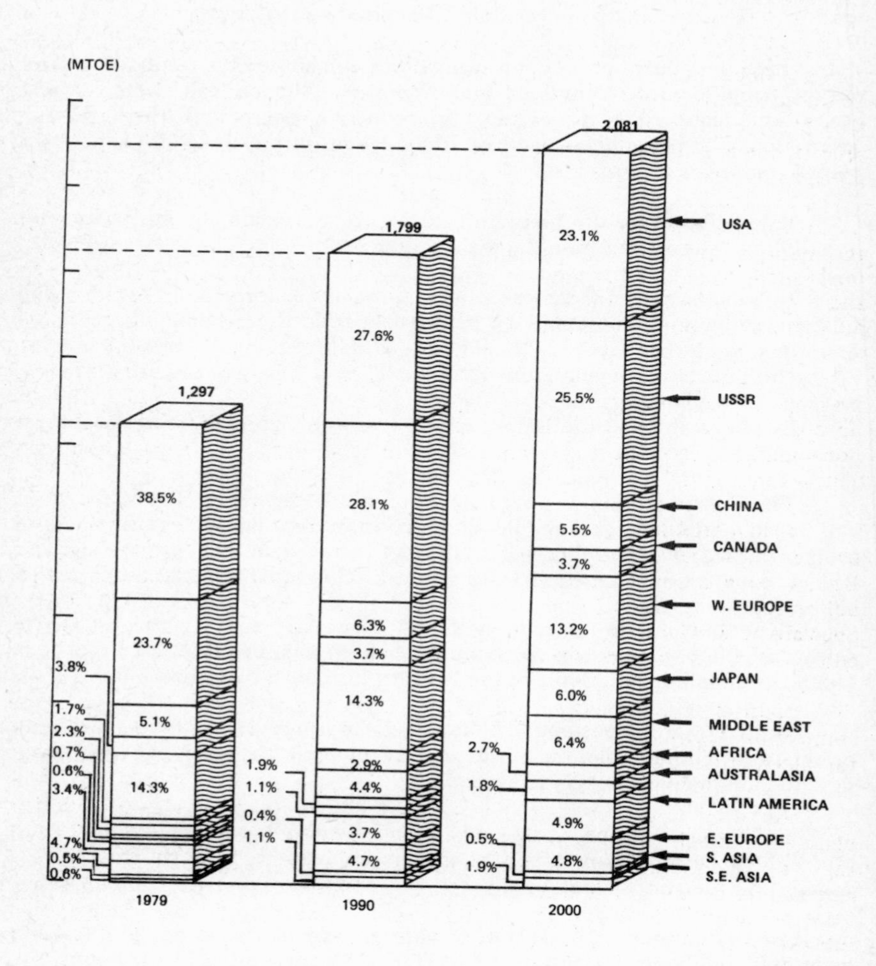

LNG (Liquefied Natural Gas).

By 1990 it is projected that the percentage traded will rise to 16% (295 MTOE) of which 40% (121 MTOE) could well be LNG, the rest being carried by international pipeline.

There are four major gas pipeline projects planned at present which could enable some 174 MTOE of gas to be traded in 1990: the 4,800 mile Alaska SouthernUSA pipeline at $23-25 billion, the most expensive civil engineering project ever undertaken, planned for total completion in 1987; the Western Siberia Europe pipeline costing $12 billion; the $2-3 billion Algeria Italy pipeline and the $2.5 billion North Sea gas gathering pipeline.

Whether or not trade in LNG will achieve the projected level depends on the nations involved carrying through supply agreements. The number of postponed, abandoned schemes to date, does not bode well for the future. At the time of writing four contracts concluded with Algeria are suspended and in all there will be a shortfall of 13.38 MTOE on contracted supplies in 1980.

The politics of energy supply is one factor which will add its looming presence to the world gas supply situation. Although the Soviet Union and Iran have the reserves to meet additional European gas requirements in the future, there may be reluctance to become too reliant on supply from these quarters. This means looking to Africa - notably Algeria and Nigeria.

LNG could be an answer to coming gas shortfalls in Western Europe. The problem however is the cost of facilities. For example, the cost of Nigeria's Bonny liquefaction plant and export terminal, with a capacity of about 16 billion cubic metres a year, was estimated, in 1979, to be $4.9 billion. Specialised LNG carriers can cost about $150 million each while a 5 billion cubic metres a year import terminal is priced at around $300 million.

Natural gas pipelines are also very expensive, although the one connecting Algeria and Italy under the Mediterranean is proving less expensive for Algeria than LNG facilities since most of the costs are met by the customer.

At the time of writing the progress of LNG is in a particularly troubled phase: Sonatrach, the Algerian state energy corportion, confirmed in late 1980 that the Arzen-3 gas liquefaction plant will not be built in 1984. This cancels agreements under which the Netherlands and West Germany were to buy a total of nearly 200 MTOE of natural gas over the next 20 years. This move followed Sonatrach's attempts to impose higher prices and a general disenchantment with LNG projects.

The Netherlands has been trying to conserve indigenous reserves by meeting supply contracts using imported gas and will probably look more to Nigerian LNG. West Germany will have to carefully consider the possibility of extra supply from Norway and the USSR to meet its own rising gas demand.

For countries like West Germany, and most other European countries, which are heavily dependent on natural gas and which have sophisticated and well developed gas transmission systems, gas will be vital for the next twenty years and supplies will be acquired at whatever cost the world market conditions fix. In Europe the pattern of gas usage to the end of the century is part of the blueprint for reduced oil dependence: gas will have to supply many of the households and other small consumers, previously supplied by oil. By 2000 about half of gas consumption will be accounted for in this way (as against 39% in 1978).

There is little alternative to this as increased coal use and nuclear power will be concentrated in electricity generation and industry and the widespread adaption of either of these to supply heat for domestic and small consumers would require far too great an expense on new supply infrastructure.

Despite a projected 60% increase in world consumption in the years to 2000, natural gas's share of world energy supply will fall from 19% in 1979 to 17% in 2000. The reserves are more than adequate to allow the expansion in production necessary and more, but the limiting factor is the expense and difficulty of the transportation and suppliers lead times for the construction of the plant and equipment necessary to import large quantities of LNG which average 8-10 years while for pipelines lead times are also long and costs high.

These are the main obstacles to increase natural gas consumption in the medium term, rather than shortage of world supply, and consumption is likely to increase well into the next century as a world gas trade infrastructure becomes well established.

Chapter 3

Oil Reserves

Since Edwin Drake drilled the first successful oil well in Pennsylvania in 1859 the world has consumed about 60 billion tonnes of oil. It is a view accepted within some quarters of the oil industry that the world may originally have possessed some 272 billion tonnes of recoverable oil, not including oil left in the ground after production or the large amounts of oil locked in tar sands and oil shales.

Since consumption in 1979 was some 3.1 billion tonnes, simple arithmetic would seem to indicate no immediate problem with oil resources - there being about 70 years worth of oil at present rate of consumption - but this is not the case.

The term 'proven reserves' means the oil which the companies consider they can exploit and proven reserves in 1979 stood at about 87 billion tonnes. Dividing this sum by 1979 consumption produces a very different result - 28 years of oil at the present rate of consumption but the situation is even less hopeful for two reasons.

Firstly, consumption has not reached its peak. Although the world's industrialised countries are now endeavouring, with some success, to reduce their dependence on oil and introduce alternatives, the world's developing countries will continue to use more oil as they lack either the inclination (in the case of energy exporters), the capital or the technology to do the same.

Secondly, the fact that exploitable oil exists does not necessarily mean that its exploitation will keep pace with world demand. Most of the world's exportable reserves are under the control of the countries who form the Organisation of Petroleum Exporting Countreis (OPEC) and the past 10 years have given ample proof that neither oil supplies nor oil prices can be guaranteed.

Examining the world oil reserve picture since 1970 it is disquieting to observe that for some time proven reserves have been diminishing and proven reserves were 2.2 billion tonnes lower in 1979 than in 1975. This means that despite the North Sea, despite Mexico, despite Alaska and despite a record level of worldwide exploration for new oil supplies, the world is using up more oil

Chart 11 LOCATION OF PROVED CRUDE OIL RESERVES JANUARY 1ST, 1980

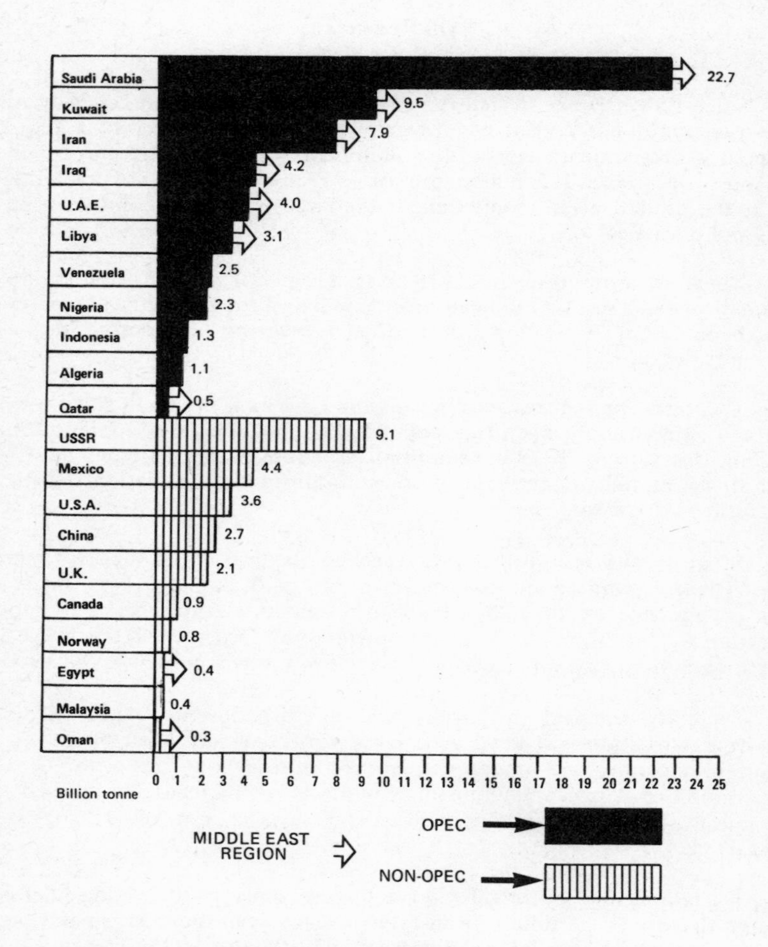

Source : Oil and Gas Journal / Own Calculations

than it is discovering.

Oil is becoming harder to find. The easy oil has been discovered and new discoveries are now coming from more difficult regions, for instance, from thousands of feet below the seabed and from beneath permafrost in some of the world's harshest regions. Even the most optimistic of energy pundits could not but agree that a point of no return has been passed with oil and to maintain supplies will become more and more a case of running faster to stand still.

In 1979 56% of the world's proven oil reserves were to be found in the Middle East, while 67% were controlled by OPEC. By contrast, in 1970, North America accounted for 8% of proven world oil reserves; in 1979 the percentage had fallen to 5%. Despite a startlingly quick growth in reserves due to the North Sea, Western Europe only accounts for 4% of world reserves and in these figures - 67% OPEC and 9% North America and Western Europe, resides the reason for many of today's energy problems: the disparity between local use and local resources of oil.

Of course there is more oil in the world than proven reserves. Two reputable sources, Halbouty and Moody and M. King Hubbert, estimate total reserves at 256 billion tonnes and 272 billion tonnes respectively while Ray Dafter, the Financial Times energy editor in his book, 'Scraping the Barrel' puts the figure far higher. According to Mr. Dafter, on top of proven conventional reserves of 88.4 BT use of enhanced oil recovery techniques will render a further 285.6 BT. Adding on a further 81.6 BT from tar sands and shale the book predicts that oil production will rise from its present level of 3,087.6 MT per year to 4,482.6 MT and maintain this level into the 21st century. Production will begin to fall below present levels by 2040 and tail off towards 2150

The opinion adopted by this study is that world production will peak during the 1990's at a level of about 4.2 BT and fall away rapidly after 2020 as alternative sources become more economical than raising small quantities of oil from the worst corners of the world and the use of expensive enhanced recovery techniques.

But speculative resources do not fill barrels. To maintain the balance between supply and demand in the 1990's a new discovery the size of the North Sea Forties field or Prudhoe Bay (Alaska) must be discovered every year. This has not happened. What discoveries there have been have tended to be small.

Mr. Rene G. Ortiz, Secretary General of OPEC, said early in 1980 that:

Chart 12 **ONSHORE/OFFSHORE OIL RESERVES**

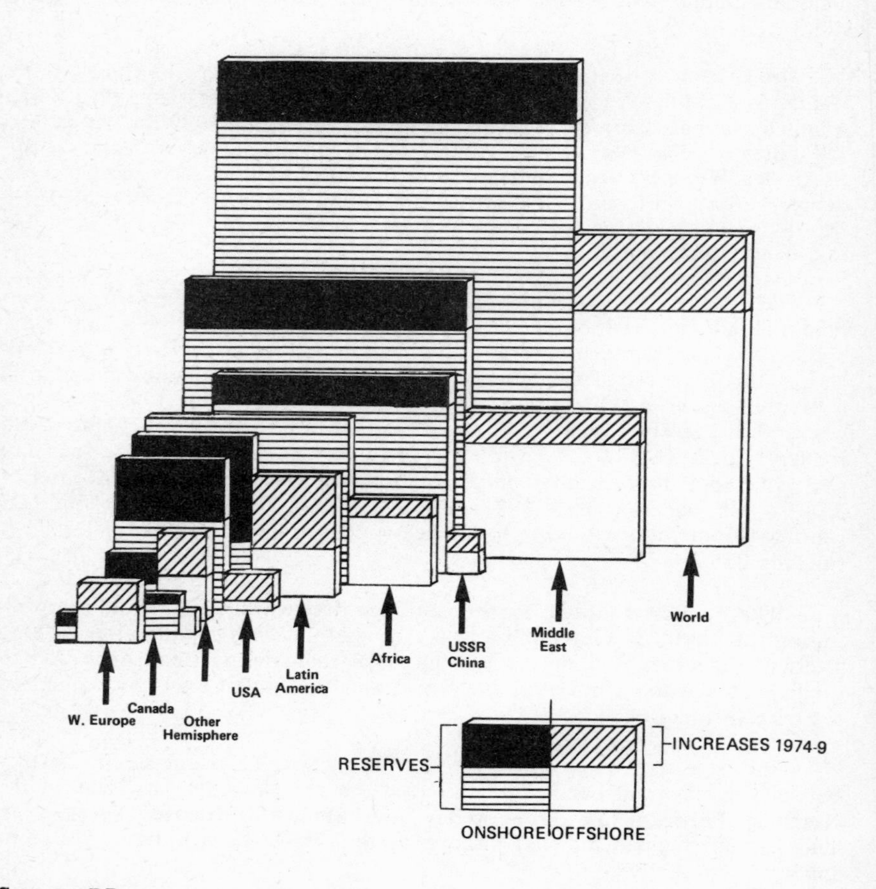

Source: BP

Chart 13 WORLD OIL DISCOVERIES IN RELATION TO PRODUCTION 1930-1979

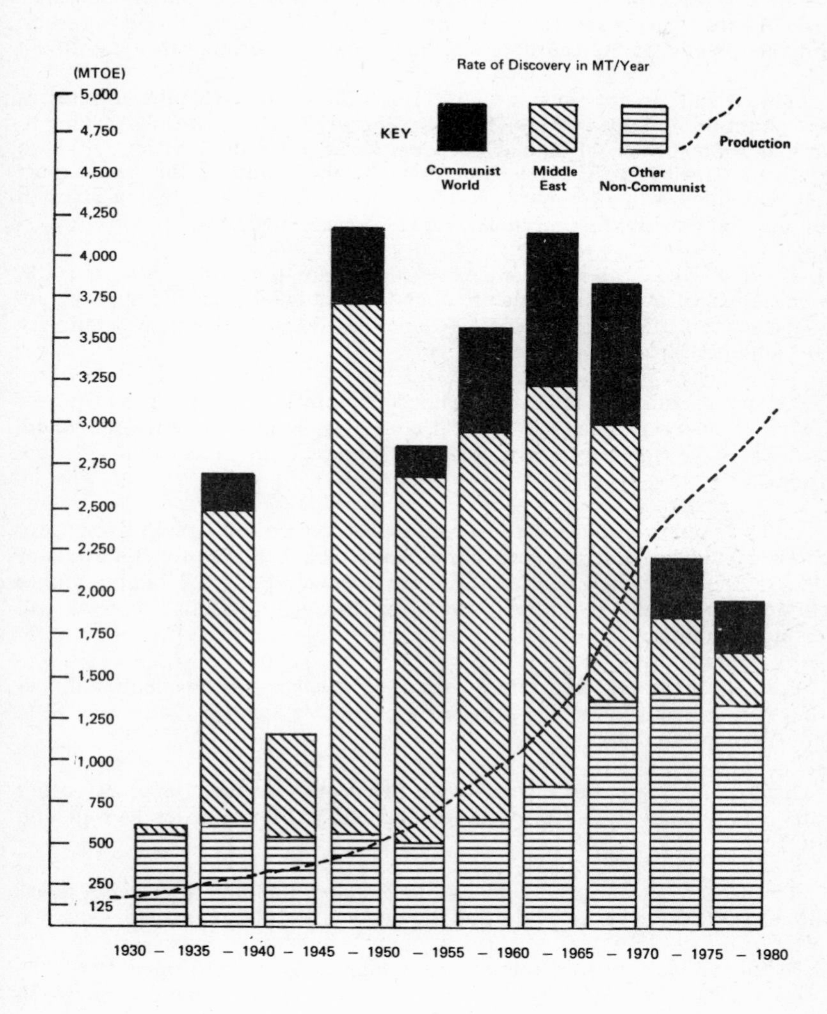

'With present world proven reserves estimated at 650 billion barrels (88.4 BT) and present consumption amounting to 22 billion barrels per year (2.99 BT) the resources will dry up within the next three decades. The present oil reserves within the OPEC member countries now stand at about 450 billion barrels (61.2 BT). If the 1980 OPEC production level (30 million barrels per day - 1.494 BT per year) were to be maintained, OPEC's output would start to decline at the end of the century, reaching exhaustion around the year 2025.'

Thus OPEC is not sanguine concerning the future of world oil reserves and exploration activity outside OPEC by countries and companies seeking to reduce dependence on OPEC has been paralleled by efforts within OPEC to maximise what, to most of the countries in the group, is the best export commodity they will ever have. Exploration for new oil is taking place in Indonesia, Gabon, Libya, Venezuela, Saudi Arabia and Nigeria.

Outside OPEC areas of intensive search are the North Sea, the US, Canada, Mexico, Australia, Argentina and Brazil while seismic surveys are taking place off the Chinese coast (China has designated energy a priority sector) and offshore and onshore in Egypt.

Oil prices and demand have enabled exploration to be carried on over much wider areas - in small third world countries, in onshore territories where conditions make for supremely difficult logistics and in isolated offshore locations.

But the point is that even if a return to the early 70's rate of discovery of about 2 billion tonnes per year (Free World) was achieved for the whole of the next 20 years, since consumption will average about 2.7 billion tonnes (excluding Communist countries) per year through to 2000, reserves will continue to diminish.

Also, to achieve such a rate of discovery when it is probable that almost all the world's really large fields in fairly excessible locations have now been found, is very, very unlikely.

Saudi Arabia has more than twice the proven oil reserves of any other country, more than the sum of the second and third countries. Kuwait and USSR.

The USSR, Mexico, the USA and China have the largest proven reserves of the non-OPEC countries while the UK lies in 13th place just below Nigeria.

Saudi Arabia's proven reserves have increased by 29% since 1970 but those of Kuwait, Iran and Iraq, the second, fourth and sixth largest reserves in the world respectively, have declined.

Proved oil reserves of both Canada and the US have declined by over 30%, demonstrating the need for transition to other energy sources in these countries.

Looking to the Far East, only Indonesia has sizeable oil resources although those of Malaysia and India have grown considerably.

Although the speculative oil reserves of USSR and China are estimated at about 25% of the world total, proven reserves in both countries have fallen over the past 10 years, those of the USSR by almost 2 BT. However, further large discoveries in both countries are likely. Russia and China are the only Communist countries with large oil reserves.

In Central and South America, proven reserves have increased considerably, mainly due to the huge new finds in Mexico, which now has the fifth largest reserves in the world, and to an upsurge in discoveries in Venezuela.

The two originally oil rich countries of Africa, Libya and Algeria have both declined in reserves since 1970, though both still possess significant oil reserves. Nigeria's reserves have almost doubled while Egypt's are declining despite significant exploration activity.

The picture for Western Europe, traditionally lacking in oil resources has been transformed by the discovery of the significant oil province on the continental shelf in the North Sea.

Oil Consumption

The degree of the disparity between where oil reserves are and where that oil is used is not only an energy problem but a formidable geopolitical problem.

It may be illustrated very simply by imagining a situation where all the world's oil trade routes closed down and all countries had to attempt to survive on their own reserves.

In 1979 North America and Western Europe accounted for 53.8% of world consumption. Western Europe's proven reserves could sustain that level of consumption for only 4.4 years, while the USA could keep going about 3 months longer. Eastern Europe is even more vulnerable at 3.4 years, and the Communist world as a whole is more vulnerable than the Free World with 19.3

Chart 14

TOTAL DISCOVERED OIL

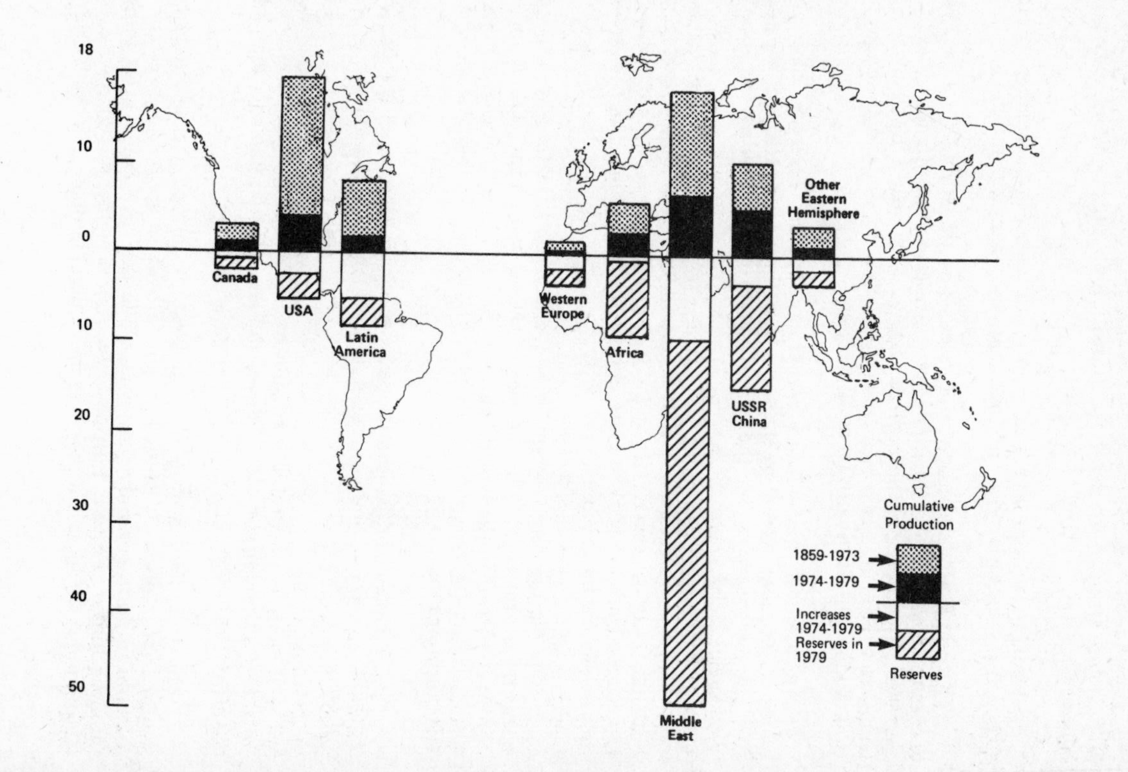

Chart 15 **WORLD OIL CONSUMPTION 1979-2000**

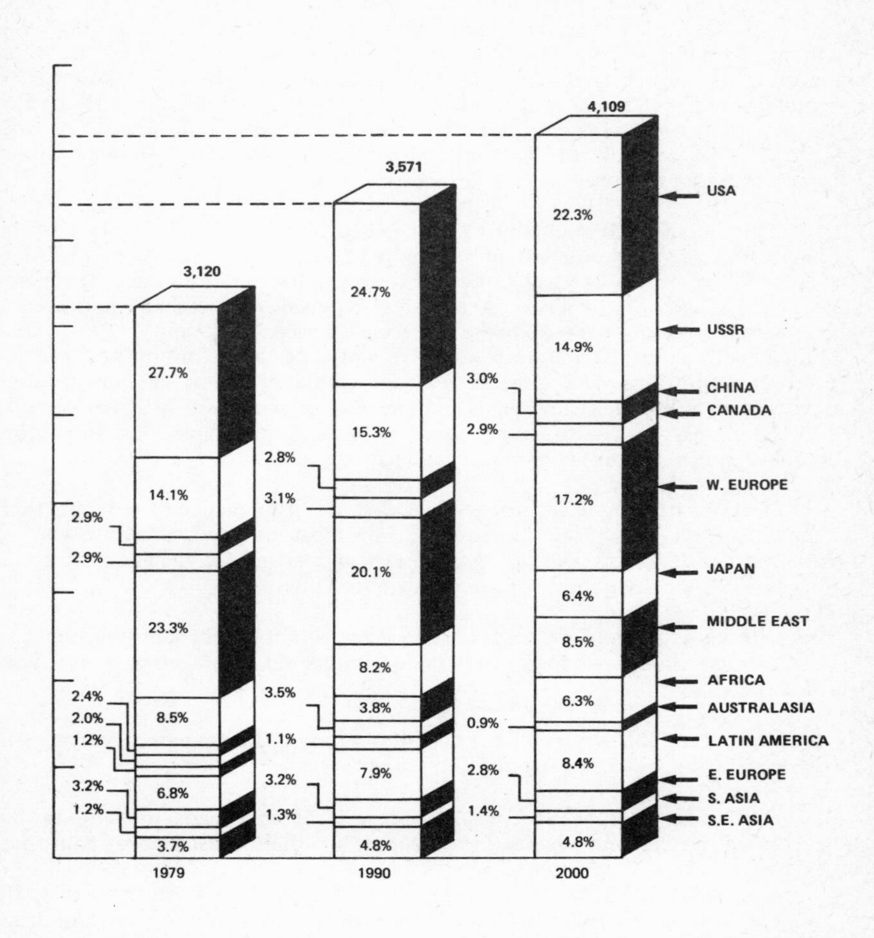

years as opposed to 30.2 years. But this is only because the Middle East's reserves fall within the non-Communist world. The Middle East could support its present rate of consumption for 656.5 years.

The point of realisation that the world's wonder-fuel has become a technological trap is long past. The world is conscious of its addiction but how to kick the habit is a different matter, only the very rich - the industrialised countries - can afford the treatment and even when the resources are there the will to act seems lacking. Over the last five years only one of the world's top ten oil users has achieved a significant drop in oil use - the U.K. - and it is no coincidence that the U.K. is also the country with the lowest economic growth. The battle to break the link between economic growth and oil usage growth has not yet been won and will not be won for some time to come without truly Draconian measures.

Oil demand will gradually be reduced in the industrialised countries - by conservation, by development of alternative fuels, by changes in energy habits such as the gradual phasing out of oil as a heating fuel and by greater efficiency but the process will be a slow one. The transition away from oil is gathering momentum and successes are being achieved. Between 1978 and 1979 the USA reduced oil use by 26 million tonnes - roughly the same amount as Sweden's total consumption. This success will be emulated in other industrialised countries over the next few years but the rate of reduction will then slow as the cheap, easy ways of using less are implemented and only the more difficult, radical and expensive ones are left.

The less developed countries (LDCs) stand little chance of reducing their dependence on oil in the short term. For most of them the only way of reducing consumption will be deprivation and economic decline. Rising oil prices could, for the poorest of them, mean just this.

For most of the LDCs increased oil consumption through the 80's and 90's is inevitable - the inseparable companion of energy intensive industrialisation.

How then will world oil consumption grow in the next twenty years and where will the growth in consumption come?

The main feature of consumption will be increasing demand in the developing world and decreasing demand in the industrialised countries.

The developing world - made up of the Middle East, Africa, Latin America, South Asia and SE Asia will increase its share of world consumption from 16.2% in 1979 to 21.3% in 1990 and 29.5% in the year 2000. The main growth area will be among the energy exporters, particularly the OPEC countries, thus the Middle East's oil demand will register an almost five fold increase (see Less Developed Countries section).

This level of growth in the Middle East and in the oil exporting countries of Africa and Latin America has serious consequences for the main oil consumers, since it diminishes the amount of oil available for export.

Least dynamic growth among the developing countries will be in South Asia where severe economic problems will hamper industrialisation - the economic problems being partly as a result of the high price of oil.

Western Europe and Japan will achieve an actual reduction in oil demand as a result of conservation and substitution of alternative fuels.

In the United States, although even more significant reductions in conventional oil consumption (i.e. oil from traditional sources), will be achieved, the growth in synthetic oil production will allow a low rate of increase in oil demand to be met. This growth in synthetic oils will be largely from oil shale (and to a lesser extent from coal). Production of synthetic oil is also included in South American and Canadian demand projections. Venezuela has significant heavy oil deposits (Orinoco heavy oil belt) and Canada's Athabasca heavy oil sands deposits will also make a significant contribution (see 'Synthetics', South America and Canada sections).

Oil Production

The last ten years have seen some big changes in the world oil production picture. Perhaps the most notable feature is the decline of United States from accounting for over 20% of world supply in 1969 to 13% in 1979. Since 1974 the USSR has been the world's largest producer of crude oil and the increase in Soviet production along with that of the Peoples' Republic of China has increased the Communist World's share of world supply from 18% in 1969 to 22% in 1979.

The Middle East also has increased its share of world crude supply, from 29% to 33% since 1969, particularly as a result of Saudi Arabia, the leading Middle Eastern producer, more than trebling production. Iraq and Abu Dhabi have also registered large increases in production but these have been balanced by the recent fall in Iranian production and the decline of Kuwait's production below 1969 levels despite an increase between 1978 and 1979.

Nigerian production has represented a powerful addition to OPEC's hold over world supply as has Indonesia's but older established OPEC members

Chart 16 WORLD OIL SUPPLY - THE OPEC CONTRIBUTION

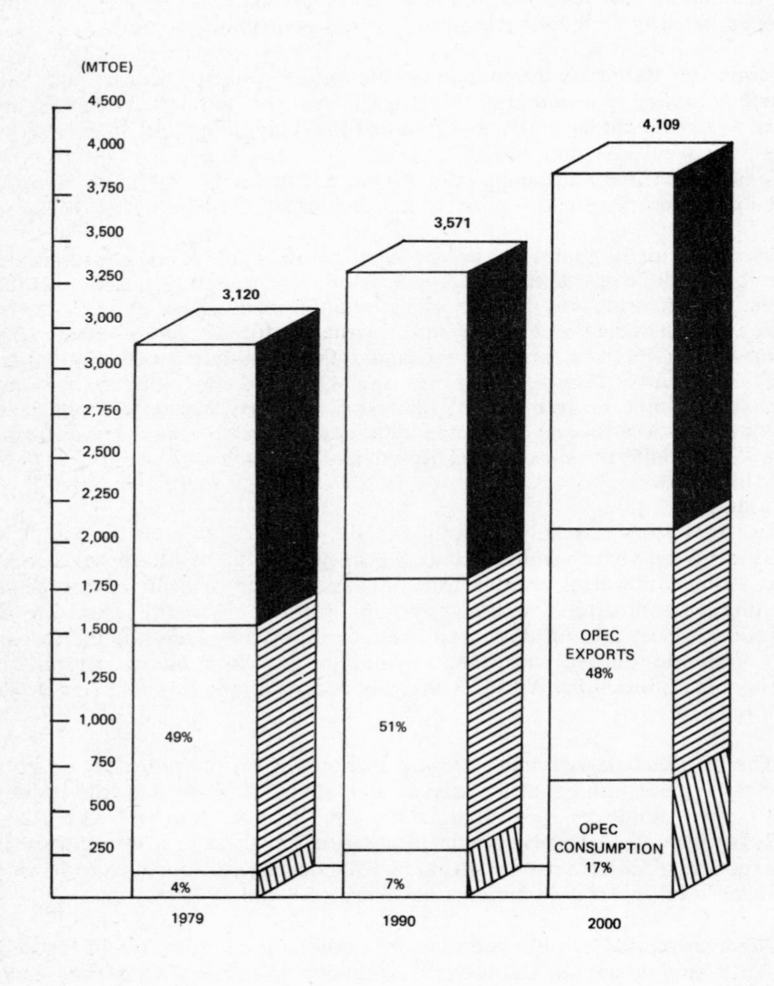

Venezuela and Libya are producing considerably less than in 1969.

Outside OPEC, China, the U.K. and Mexico have been the most significant additions to world production.

At present (December 1980) the world oil situation is, at first sight, paradoxical. In 1979 the loss of Iranian production as a result of the revolution created a worldwide supply crisis. Oil product prices (e.g. petrol) went up in many countries and there were shortages. The present war in the Gulf between Iran and Iraq has reduced supplies of oil to the importing countries by an amount equivalent to around 200 million tonnes a year out of world (non-Communist) total consumption of 2,500 million tonnes a year - an amount considerably greater than the loss in 1979. Yet the disruption of supply has not caused a world shortage, indeed the situation is one of surplus.

There are a number of reasons for this. Firstly, world oil consumption has decreased since 1979 as a result of the reduced industrial activity which characterises recession in conjunction with conservation measures and the transition, in industrialised countries, to alternative fuels. There has also been a continuing increase in supplies from non OPEC areas e.g. Mexico or the North Sea, while some OPEC countries have increased output to prevent supply disruption. This combination of circumstances has led to (non-Communist) world inventories (stocks) of oil being at their highest ever levels - high enough to make up for the Iran-Iraq production loss for a number of months before the inventories return to historically normal levels.

However such a situation cannot continue indefinitely and the fact that so significant a loss of supply has been easily maintained for a number of months should not be taken as an indication of a future supply pattern where demand keeps comfortably within an available supply. Economic upturn, in conjunction with a continuation of loss of supply from Iraq and Iran can quickly wipe out the world's oil supply surplus and even continuing modest rates of economic growth without increased supply would lead to shortfall in the winter of 1981.

The supply situation could also be changed by OPEC action to create a tighter demand position to facilitate price increase.

Looking ahead, world oil production will continue to increase to the year 2000 at a rate considerably below that recorded over the 10 years between 1969 and 1979.

All areas should actually increase production - Latin America will show a particularly large increase as a result of Mexico's expanding oil industry, and, to a lesser extent Venezuela's supplementing of conventional oil production by synthetic oils.

The United States will also be able to increase indigenous production by means of synthetics. Increases in production by USSR and China will be conventional although some manufacture of synthetic oil from coal will be carried out in USSR. The discovery of large new reserves is more likely in Russia and China than in many parts of the world, Russia because of the vast tracts of land still to be thoroughly explored in the North of the country and China because offshore exploration is still in its infancy and discoveries so far suggest good potential.

Africa's production will not increase much - although Nigeria's contribution may grow a little, this will be balanced by reductions in production from Libya and Algeria. Western Europe's North Sea production is likely to peak in the 1990's declining to a level some 20% above the 1979 level by 2000, however this pattern could be changed by Government policies restricting output or by significant new discoveries in the U.K. or Norwegian sectors.

The overwhelming conclusion, having established a likely demand and production outlook for the world is a depressing one. Despite increased production in Mexico and the North Sea; despite the development of synthetic oils in the US, Canada, South America and Australia; despite the transition to alternative fuels - coal, nuclear and gas; and despite conservation, the world's dependence for crude oil on the cartel called OPEC, will actually increase to around 1990.

In 1979 OPEC was responsible for 47.9% of world oil production. By the year 2000 that percentage is likely to have risen to 49.8% Although by virture of increased indigenous consumption, the amount of oil actually exported by OPEC will decrease slightly by the year 2000, as compared with 1979, the degree of dependence will be almost as great, with OPEC still responsible for 48% of oil movements (as against 51% in 1990 and 49% in 1979).

The fact that the oil the world needs to support even the comparatively modest level of economic growth projected in this study depends on OPEC increasing production above estimated maximum output capacity (1979) of about 1,700 million tonnes of oil, means that OPEC's oily hands will be clamped just as firmly around the world's throat for the next 20 years and beyond.

Increased dependence may also have the effect of returning oil price control to OPEC since 1974 prices have been distorted by market forces as much as by agreed OPEC decisions.

The USA's need for imported oil will peak around 1990 at some 470 million tonnes of imported oil per year and subsequently fall by about 100 million tonnes by 2000, aided by growth in the synthetics contribution. The USSR and China will continue to gradually increase production as new fields are brough on stream. Consumption will grow slightly faster than production in the USSR and slightly slower in China but both will still be net exporters in 2000. However, Eastern Europe's dependence on imports will grow.

Although Western Europe's oil import requirement in 2000 will be less than it was in 1979, the North Sea contribution will probably be declining.

Western Europe and Japan will be almost as vulnerable to OPEC supply disruptions in the year 2000 as they were in 1979. The USA will be less so due to increased indigenous production through synthetics and also through an increased proportion of supplies from Mexico.

In 1979 25% of the USA's oil imports were from the Middle East and about 30% from Latin America. These proportions will change steadily and the OPEC proportion will fall as Mexican production expands. Exports from Venezuela could fall depending on the speed with which synthetic oil production can be developed.

Europe's dependence on Middle Eastern OPEC countries - 66% in 1979 will probably be slightly greater by the year 2000 as imports from North Africa (14%) and possibly the USSR (9%) gradually decline.

Chart 17 WORLD NUCLEAR ENERGY CONSUMPTION 1979-2000

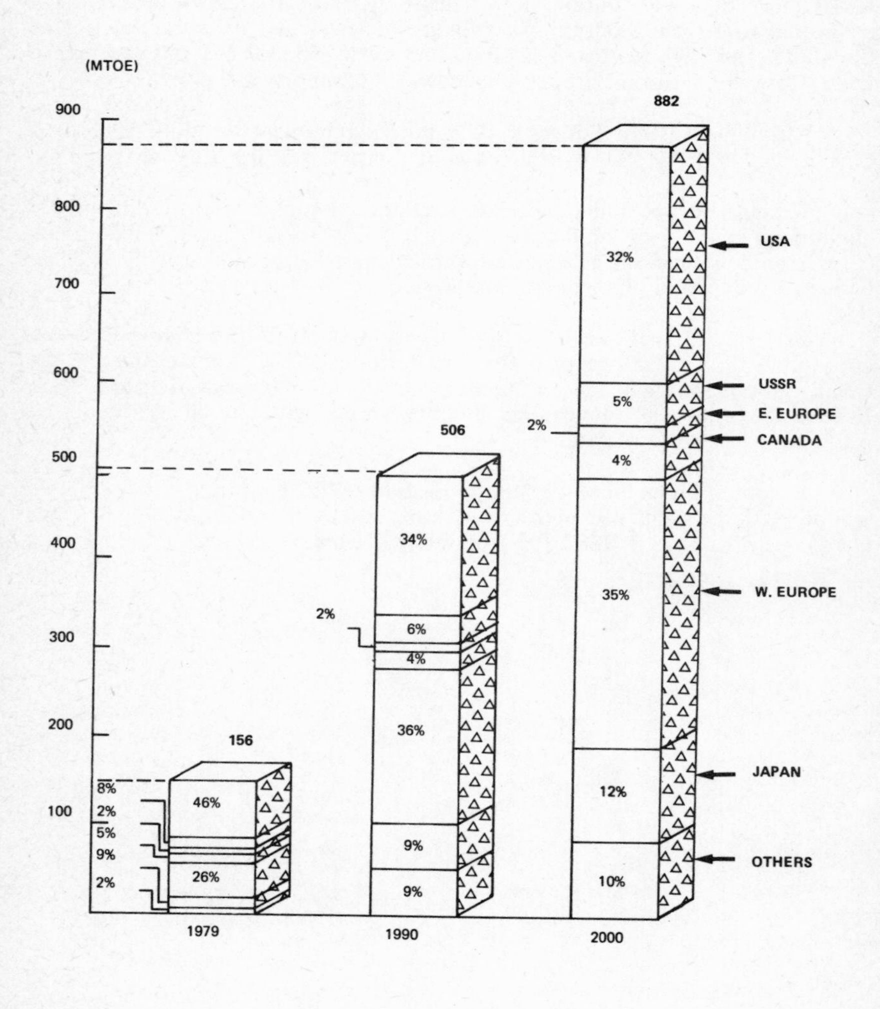

Chapter 4

Nuclear Power

The future is not what it used to be as far as nuclear power is concerned: year after year predicted capacity has fallen short of actual capacity by amounts in direct proportion to the length of the prediction. Projections of OECD nuclear generating capacity are a case in point, in 1970 the prediction for 1985 was for 563 gigawatts (a large nuclear power station generally has a capacity of about 1.2 gigawatts) the projection in December 1978 was for 214 gigawatts. But why are nuclear power stations the late 20th century equivalent of castles in the air?

After all, nuclear power is still regarded, if not as the energy panacea which its supporters styled it in the 1950's, then at least as one of the cornerstones of any rational world energy strategy aimed at reducing dependence on oil. At the World Energy Conference in Munich in 1980 and at the Fifth Annual Symposium of the Uranium Institute in London, Dr. Ulf Lantzke, Executive Director of the International Energy Agency said that nuclear generating capacity must increase fivefold by the end of the century if the industrialised world is to achieve even a modest economic growth rate.

Forecasts made at the World Energy Conference envisage almost 13% of total world energy production accounted for by nuclear power in 2000 and 31% in 2020 but it is difficult to view such projections as much more than wishful thinking, involving as it would an increase of some 1000 gigawatts above today's installed capacity. With only around 250 gigawatts on order at present and lead times of around 7-10 years for construction a huge upsurge in orders would be required, to come anywhere near the eightfold increase in capacity necessary to reach such a level.

Optimistic forecasts in either the near or medium terms seem even less justifiable when one considers that the number of cancellations is higher than the number of new reactors ordered for the third consecutive year. In all, some 34 reactors with a total capacity of 28.5 gigawatts have been commissioned in the Western World over the past three years; 36 plants (34.4 gigawatts) were ordered but 48 projects (50 gigawatts) were cancelled, of which 32 were in the United States.

The present depressed state of the nuclear industry is further exemplified in the fall in the spot price of uranium in spite of the fall in output of mines and mills.

What then is going on? Nuclear fuel, uranium, is at present in plentiful supply and it is generally true that nuclear fission is the cheapest way of generating electricity.

The trouble is that nuclear power is the most political of energy issues, the field most charged with unpleasant association and fears. Nuclear power has caused changes of government, stimulated huge public demonstrations and provoked breakdowns in international relations.

One of the central problems is how to expand nuclear power without simultaneously precipitating the spread of nuclear weapons. The USA controlled the early development of nuclear power in the Western World, supplying the technology and materials for most countries' nuclear construction programmes. However, US confidence in international non proliferation measures was destroyed by the Indian nuclear explosion of 1974 followed by the French and German agreements in 1975-6 to supply nuclear technologies to Third World countries such as Brazil, Pakistan and South Korea. Rigorous new restrictions were imposed on trade in nuclear materials and development while countries refusing to comply were threatened with loss of enriched uranium supplies, the fabrication of which was monopolised by the USA.

In the early seventies however, this monopoly began to be eaten away and France and Germany, having rendered themselves largely technologically independent of the US, began an aggressive export drive to Third World countries, some of which were widely regarded as having military motives for their nuclear programmes.

In the later seventies, international tensions created by the proliferation issue have eased somewhat with the nuclear exporters, Germany and France, becoming more conscious of the dangers and the US moderating its 'Pax Americana' stance.

The International Nuclear Fuel Cycle Evaluation (INFCE) which President Carter set up in 1977 to study the technical aspects of his anti proliferation proposals, is partly responsible for the improved relations, providing as it did, a forum within which the 46 participating countries could air their views. However, the main reason for the easing of tensions is probably the swift decline in the international nuclear business; less nuclear reactors sold means less potential proliferation.

However, the calm is likely to be only temporary - the end of the calming period of INFCE plus the expected upturn in orders will bring the various points of contention to the surface again. One of these is the issue of reprocessing and of fast breeder reactors. The US has, in the past, been opposed to the commercial development of these uranium saving technologies on the grounds that they make plutonium - which is used to fabricate atomic bombs - too easily available.

Europe and Japan disagreed strongly with the US,believing that, with the spectre of uranium scarcity a constant possiblity, they could not afford to deny themselves what are essentially conservation technologies.

Whether this line of argument will hold up in the early eighties is open to question - there being no shortage or potential shortage of uranium, for at least the next decade.

The questioning by Governments of the wisdom of supplying the means to make atomic bombs to countries considered politically immature is however, only part of a wider problem for the nuclear industry, referred to by Dr. Ulf Lantzke as 'the paralysing crisis of confidence in the future of nuclear power'.

Anti nuclear protest has had mass support in some countries. In the USA, which has experienced the heaviest nuclear slowdown, the anti nuclear movement has identified nuclear power with big business. The Three Mile Island debacle proved the largest single public relations disaster in the history of energy supply, while the popularity of the film 'The China Syndrome' may be viewed as providing a focus for the fears of nuclear power harboured by many people.

Anti nuclear groups rally around the following: firstly, the fact that nuclear power is connected with nuclear weapons; secondly, the safety problems as exemplified in the pattern of accidents at the pressurised water reactor at the Three Mile Island Plant near Harrisburg, Pennsylvania; thirdly, the problem of radio-active waste. In addition anti nuclear groups generally advocate a return to less centralised energy production, making use of renewable sources such as solar, wind and wave-power.

Anti nuclear groups and the amount of public and political support they can command have proved a considerable factor in nuclear industry's decline. Opposition has proved strong enough to block all ideas for new reactors in Holland, Norway, Denmark and Austria. There have also been major demonstrations in Spain, West Germany and the U.K. Only France, with the most ambitious programme of all for nuclear development has been comparatively free of effective opposition. It is perhaps, of interest to observe in passing that the differing impact of public oppostion on nuclear projects is directly related to political structures in the countries concerned i.e. the existence of channels through which public dissent may be transformed into political action.

In countries where detailed justification of nuclear power is required immediate necessity tends to be much more difficult to put over than long term desirability.

Although nuclear power is easy to incorporate into a long term energy plan, building vastly expensive power stations (the two ordered in the U.K. in 1980 at Heysham and Torness will cost £1.2 billion each) when electricity demand is falling is another matter.

The economics of nuclear power has been a long running debate. Although it is perhaps now generally accepted that nuclear generated electricity is the cheapest - the greater cost of nuclear plants than fossil fuelled plants leads to accusations of undue optimism in the compilation of cost per kilowatt statistics pertaining to incorporate all capital costs incurred. Also public mistrust of nuclear power has not been helped by the constant inaccuracy of Government predictions of future levels of usage in many countries.

Nuclear Power to the Year 2000

All these problems may be held partly responsible for the poor performance of nuclear power over the next 20 years. Although worldwide, the increase from the 1979 level of 156 million tonnes of oil equivalent (only 2% of world energy consumption) to 880 MTOE in 2000 (7% of world energy consumption) makes nuclear power the world's main growth fuel, the paucity of the contribution compared to that of oil (33% of world total in 2000) testifies to the world's poor energy planning. Put simply, blinkered, short term economic planning and fear are causing us to waste vital liquid hydrocarbons, by burning them up, while in effect conserving uranium, a substance which has no useful application other than producing heat in a power station.

The main growth areas for nuclear power during the next twenty years will be Western Europe and Japan. In Western Europe, France and West Germany will be the chief growth areas.

Japan is the only one of the leading industrialised nations more dependent on imported fuel than France and therein lies the particular reason for expansion in these two countries. Necessity is likely to be the final impulse to move the world into a more nuclear age.

The USA and Canada have in their burgeoning synthetics industries an alternative to fullscale nuclear expansion, which does not arouse the same public antipathy.

The USSR could easily expand its nuclear capability by more than the modest amount (from 12.5 MTOE in 1979 to 47 MTOE in 2000) predicted here but will probably place more emphasis on its vast hydrocarbon reserves in the short term and the resources available from their speedy exploitation and export while pushing ahead with the development of Fast Breeder Reactor technology and the more ambitious nuclear fusion.

Uranium

Had it not been for the downturn in the nuclear industry the world might have been suffering a uranium supply crisis in the early 1990's. As it is, this will not occur but the present state of reserves versus consumption makes nuclear power superficially the most vulnerable of the primary fuels and a very poor bet to reduce world oil dependence. At the rate of consumption which seems likely in view of projected increases in world (non-Communist) generating capacity, by the year 2000, 82% of the world's reasonably assured uranium reserves will be gone. This depressing statistic would make nuclear power a blind alley.

Uranium consumption in 1980 will be about 30,000 tonnes - well below maximum attainable production capacity which is 50,000 tonnes.

Up to around 1990, predicting the amount of uranium which will be used is fairly straightforward since it may be based on the number of reactors operating and those under construction. This simple relating of operational capacity and uranium consumption is possible as long as the recycling of uranium and plutonium do not make a significant contribution and while fast breeder reactors are not in major use.

Thus, working to a predicted level of installed capacity of 334 gigawatts in 1990 (Free World) - a figure which is lower than the lowest point in the unrealistic range suggested by the INFCE - it is reasonable to assume that about 90,000 tonnes of uranium will be consumed in 1990.

Beyond 1990, prediction is more difficult. INFCE postulates 200,000 tonnes of uranium consumed in the year 2000 if there is high nuclear power growth a preponderance of light water reactors and no increase in efficiency or recycling. The opposite case would be for growth with recycling of uranium and plutonium and their use in fast breeder reactors.

This study predicts low growth in conjunction with little recycling and little use of the fast breeder reactor (FBR). These would seem to be the most likely combination of circumstances at the present time. Low growth in capacity is likely for the reasons already outlined. Recycling will probably not grow until it has to because it is controversial. The fast breeder will make little contribution this century; firstly, because its technology makes it seem more of a proliferation risk and secondly because it is 50-80% more expensive than conventional thermal reactors. The FBR will not be significantly used until growing need for nuclear power, the beginnings of uranium scarcity and greater confidence in the technology make it inevitable, but this is likely to be well into the next century.

The USA's go-slow on the fast reactor is unlikely to be reversed for a number of years because of nuclear power's poor public image in the US. Also, the present impetus in the US is towards synthetics which has the advantage of being more directly substitutable for oil than nuclear power and does not have a Three Mile Island to contend with. Following the low nuclear power growth, low uranium conservation scenario, consumption in the Free World will probably reach some 170,000 tonnes by 2000.

This will mean that around 2.1 million tonnes of uranium will have been used between 1980 and 2000, or 82% of the 1979 reasonably assured reserves.

In fact the situation is not as bad as this. The reasonably assured uranium resources already mentioned are made up of uranium which may be mined at a cost of up to $130 per kilogram. Estimated additional resources at the beginning of January 1979 make speculative reserves some 5 million tonnes but this could, were demand sufficient, be supplemented by less accessible, more expensive to produce uranium, should this be necessary, in the same way as offshore oil, once considered prohibitively expensive to produce, is now highly profitable.

In addition, uranium is by no means an uncommon resource and, compared with oil, gas and coal, very little time and money have been devoted to its discovery. Reserves are at present at their highest level and this trend is likely to continue. Known reserves have expanded significantly in recent years due to significant new discoveries in Brazil and Canada and improved knowledge of deposits in the Central African Republic, Namibia, South Africa and the United States of America.

Potential uranium exporters are the countries well endowed with reserves exploitable at below $80 per kilogram. The United States has the largest (Free

World) deposits. However, Australia, the next largest deposit of easily available uranium has no nuclear programme and will be, with South Africa, Canada and Namibia and Niger, one of the main exporters.

At present, there are exploration programmes in some 40 countries with exploration expenditure running at around £300 million per year, an extremely low level of expenditure when compared with that spent on oil exploration. New mines scheduled to begin production in the 1980's and expansion of production in the US, Canada and Australia will mean that production will keep ahead of demand through the 1980's. In the 1990's, production in smaller, less developed countries will become more important.

After the year 2000, production from present sources will fall below consumption and new demand must then be met by new discoveries. This means that if nuclear power is to be the major source of energy envisaged by the World Energy Conference, supplying more of the world's energy needs than oil in the year 2020, a vast increase in present level of exploration is required now. But finding more uranium is not enough if nuclear power is to be anything more than an expensive breathing space. A change in technology is also required to the reactors which do not deplete resources at such a rate.

Type of Reactors - The Future

At present the Free World has some 150 gigawatts of nuclear generating capacity. Apart from experimental FBR's (Fast Breeder Reactors), 0.3% of world capacity, these are all thermal reactors. A thermal reactor is one in which the chain reaction (a neutron induces a nucleus to fission causing neutrons to be released which cause further fissions) is sustained primarily by fission caused by thermal neutrons, (neutrons which have been slowed down by a moderator so as to make them more likely to collide with a fissile nucleus and cause more fission). These reactors use a moderator to slow down the neutrons produced in fission.

The various thermal nuclear types of reactors are basically differentiated by their different types of moderator.

The most common type is the LWR (Light Water Reactor) - so called to differentiate if from the HWR,(Heavy Water Reactor). In 1980 some 87% of the world's capacity was accounted for by this kind of reactor. In LWR's (or PWR's - pressurised water reactors) water is both the moderator and the coolant. The water is kept under pressure in a closed, primary loop to prevent boiling and is circulated through a heat exhanger which generates steam in a secondary loop connected to the turbine. This was the type of reactor at Three Mile Island.

Accounting for about 5.2% of world capacity is the other type of water moderated reactor the HWR (Heavy Water Reactor). The most widely used design of HWR is the Candu (name derived from Canada, Deuterium and Uranium). The moderator here is heavy water - which is water containing a substantial proportion of deuterium atoms.

Other types of reactors are: the GG (graphite gas) reactor, those that use graphite as a moderator instead of water (4.9% of world capacity in 1980; the AGR (Advanced Gas Cooled) - 2.0% of world capacity. Both of these use gas, usually carbon dioxide, as a coolant instead of water and graphite as the moderator. A variant of this type is the HTR (High Temperature Gas Cooled Reactor). This type of reactor, together with the BWR (Boiling Water Reactor) account for only 0.5% of world generating capacity between them.

In a recent survey of nuclear plant performance using load factor as an indication of the reliability (i.e. the percentage of time during a year that a reactor has run at full load) the Candu HWR achieved the best load factor in 1976, 1977 and 1978 followed in second place by the GCR (Gas Cooled Reactor) in 1976 and the PWR in both 1977 and 1978.

Looking ahead to 2000, the INFCE predicts that the LWR (PWR) will still dominate nuclear power at 86% of Free World capacity with the HWR (Heavy Water Reactor) achieving significant growth to 8.8% of total and the share of the AGR falling.

Although a large percentage increase is predicted for the FBR (Fast Breeder Reactor) by 2000 its total share of nuclear generation is only expected to reach 2.6% or 21.8 gigawatts and this may be regarded as optimistic.

The FBR is not a thermal reactor. This means that it does not use a moderator to slow down the neutrons. The fission chain reaction in an FBR is sustained by fast neutrons (neutrons that have lost little energy since being produced in the fission process). The great advantage of the FBR is that fertile uranium 238 is placed around the core and this is converted into plutonium. (FBR's are fuelled by a mixture of plutonium and uranium oxides.)

Current thermal reactors use only about 1% or less of mined uranium (uranium is a mixture of the isotopes U-235 (0.7%) which is fissile, and U-238 (99.3%) which is fertile - fertile means that it can be converted into a fissile material by neutron capture). With breeder reactors about 50% of natural uranium can be used. This means that utilisation factor of uranium reserves is vastly increased.

It would, in fact, mean that present proven uranium reserves would contain more energy than proven coal, oil, gas, oil shale and tar sands reserves combined.

However, there are a number of problems which suggest that the FBR will be very slow to make a significant impression on world energy supply.

Firstly, the economics of the fast breeder (and of reprocessing) are based on the expectation of a rise in uranium prices. The British Department of Energy (Energy Paper, 39) estimated that early (the first 10 or so) British fast reactors would have capital costs some 40-80% above the already very expensive thermal reactors. Further, although fuel cycle costs would probably be cheaper, even so the fast reactor would not compete economically unless uranium prices were somewhere between $175 and $330 per kilogram, which is between twice and four times the present level.

It has generally been expected that once increased nuclear capacity begins to take up the present uranium glut prices will rise. Some recent thinking (e.g. by SPRU - Sussex University Science Policy Research Unit) has suggested that lower expectations of capacity, coupled with new discoveries in Canada and Australia of ore grades considerably cheaper and easier to mine (for example, than those in the US) plus technical advances in utilisation mean that unless an OPEC like supply cartel emerges, prices will probably not rise significantly this century. This would put the economic day of the FBR some 40 or 50 years in the future. The corollary of this would be a likely shift in Research and Development policy away from uranium conserving technologies and towards improvements in thermal reactor design and in techniques for uranium discovery.

Chart 18 FUEL USED FOR ELECTRICITY GENERATION 1979-2000 (excluding Africa, Middle East, China, E. Europe, USSR)

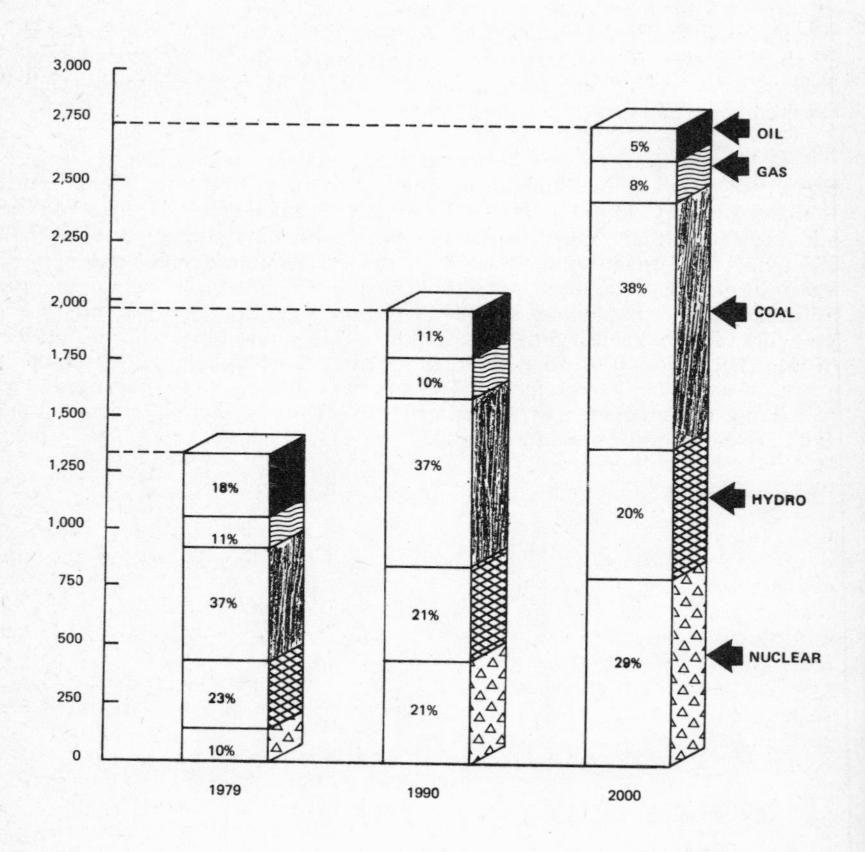

Chapter 5

Electricity

If we look far enough into the future, there is no doubt that electricity will be the main means by which energy will be concentrated into usable form. When renewable energy sources finally displace fossil fuels it will be to electricity that the energy of sun, wind and waves and geothermal sources will mainly be converted.

The problem with using any fuel to make electricity is the poor rate of conversion. When coal, oil, gas or nuclear power are used to produce the heat that raises the steam that drives the turbine that operates the turbo generator, almost two thirds of the energy contained in the original fuel is dissipated in the process. When renewable sources of energy such as wind, wave and solar power are used this will not matter because the amounts of energy available are, for all practical purposes infinite, however, when it comes to fuels which are both finite and more useful when used for purposes other than producing heat,poor conversion equals waste.

It is a sad fact that the worst possible thing to do with our oil, gas and coal resources (apart from dropping them into the sea) is to burn them, whether under the boilers of power stations or under the boilers of domestic or factory heating systems.

Oil, gas and coal are fossil fuels and as such they will run out. Oil - crude petroleum - contains a variety of hydrocarbons from dissolved gases to gasolines and kerosines to bitumen and waxes. It is the most valuable of the fossil fuels because it is the easiest to obtain, transport and use. It is most indispensable as a chemical feedstock and as a transport fuel and least indispensable as a heating fuel. Gas and coal will become our sources of hydrocarbon feedstock when oil runs out (along with synthetic oils from oil shales and tar sands). Oil at present accounts for about 18% of electricity generation*, gas for 11% and coal for 37%. Thus fossil fuels between them constitute nearly 70% of the fuel we use to make electricity.

By the year 2000, this will have fallen to just over 50% as a result of the

* all percentages refer to the world excluding Africa, Middle East and Communist countries.

increase in electricity generation by means of hydro and nuclear power.

Although the rate of increase in electricity use will be well below the level of 7.5% per annum between 1963 and 1973 and 5.2% per annum between 1970 and 1978, growth will continue at rates of around 2% in industrialised countries and higher rates in non-industrialised countries.

In 2000, the dominant fuel in power stations will be coal, accounting for nearly 40% of the world's electricity generation. The level of coal usage will in particular areas vary greatly being largely dependent on indigenous resources, despite the increase in the world coal trade. Thus in Australia, by the year 2000, coal will account for 84% of electricity generation. In another of the chief coal producing areas, North America, the proportion will be 48%.

Although actual use of gas to generate electricity will increase by some 50% between 1979 and 2000, as a proportion of total electricity generation fuel used, it will decrease from 11% to 8%.

Coal, on the other hand will increase its share of electricity generation to nearly 40%. Two factors make coal a more rational fuel for electricity generation than oil or gas. Firstly, reserves are not, in the near future, a problem, secondly use of coal as a feedstock for synthetic transport fuels manufacture, for SNG (substitute natural gas) manufacture and for petrochemicals is a more expensive process than is the use of oil and gas for these purposes. Although manufacture of synthetic fuels from coal will be extensively and increasingly practised, its most economical use will continue to be as a heat producer, until the situation is changed by advances in nuclear generation and chronic hydrocarbon shortage.

Oil will only account for about 5% of electricity generation in 2000, a fall of 40% below the 1979 level.

Oil use to generate electricity will however, increase in the oil exporting countries, which lack the economic incentive to change to alternatives.

By 2000, nuclear power will account for nearly 30% of electricity generation. Unlike the fossil fuels, which have a multitude of uses, uranium has only one. Thus considerations such as (i) the poor conversion of fuels into electricity and (ii) depletion, are less important*. (Although this is not to say that improvements should not be made in efficiency of nuclear generation and in efforts to eke out nuclear fuels by means of the FBR - changeover to the FBR is vital.)

* in fact the same physical laws do not apply to the conversion of uranium into heat.

Although electricity generation by renewable means will increase by 75% 1979-2000, this will be lower than the general increase and the 'renewable share' will fall. Most of the 75% increase will be accounted for by new hydro electric schemes, many of which will be in South America although a small amount will be contributed by wind, wave, geothermal etc.

When examining the likely changes in the pattern of electricity generation it should be remembered that this is not simply a case of fuel input and electricity output. Electricity production must accord with demand as closely as possible, otherwise waste of fuel or blackouts will occur. Sophisticated planning techniques enable generation in the industrialised countries, generally, to follow the peaks in day to day and seasonal usage with reasonable accuracy. However, a certain amount of 'on line' generating capacity is kept in reserve ('spinning reserve') for unexpected demand surges. This spare 'on-line' capacity must be kept to a minimum or massive fuel wastage results and to this end a certain amount of peak load shaving plant must be in constant readiness. Peak load shaving plant means small power stations - able to deliver power into the grid very quickly reaching full capacity in a matter of minutes, or even seconds of start up. These plants are generally gas turbine power stations. Although expensive to run - each kilowatt generated being far more expensive than a kilowatt generated by a large nuclear or coal fired station - their very presence on the system is a large cost saver.

The necessity of such stations means that there will always be a minimum level of oil (or gas) fired plant on any system.

However, the problem of expensive-to-run, peak load shaving plant, and wasteful 'over-generation' can be mitigated to a great extent in two ways.

Pumped storage stations are one method. Basically these act as gigantic batteries soaking up surplus generation. Electricity generated when demand falls below predicted level is used to pump water from a low to a high reservoir. This water may then be returned to the lower reservoir through a hydroelectric turbogenerator to meet sudden peak demands. Although more energy is used in raising the water than is generated by its fall, these pumped storage stations are energy savers since they reduce the need for the construction of conventional thermal plant which would operate at a low load factor.

The second method is interconnection. The electricity supply system in any country must be geared to follow peaks and troughs of demand. Generally, highest demand comes in periods of cold weather at times of day when many electrical appliances are being simultaneously used - the end of a popular television programme in Britain generally signals a sudden demand increase as millions of people make cups of tea. Generating plant operates most efficiently at full load which means that the more electricity is produced by stations operating at full load, the cheaper that electricity will be to produce. Power stations themselves vary in efficiency - nuclear stations are the cheapest generator of electricity when comparing stations at full load and engineers at central control points operate a merit order of stations in order to ensure that, at any given time, the most efficient combination of stations is on line.

Interconnection, which means linking the electricity systems of two countries, is a method of fuel and money saving. The idea behind it is that since day to day peaks in electricity generation will rarely fall at the same times, peaks in one country may be met by high merit plant generated electricity (i.e. cheap electricity) from the other and vice versa. There are already numerous interconnections across Europe. A low capacity direct current link joins the French and British systems and a larger one (2000 MW) is planned.

Fuel used for electricity generation will increase from just under 30% of total fuel use to 37% by 2000. Although gradually slowing due to the saturated markets in the industrialised world for many electrical appliances, electricity generation will continue to take a larger share of the fuel the world uses. This is because, despite conversion losses, electricity will have an increasing number of unique functions in a world becoming more technologically sophisticated. Also it is the best way of transporting and distributing energy and is almost infinitely flexible in its possible uses.

A train bringing coal from one of Russia's huge coal fields in Siberia to Moscow would use up most of its cargo on the journey. How much easier to build power stations by the mines, then send the coal to Europe by wire.

The inefficiency of the energy conversion process from fossil fuel to electricity may be improved from its present level of 30-35% using turbo-generators, to an efficiency of 50% by the use of MHD (magneto-hydrodynamic) power generation. MHD is the direct conversion of heat into electrical energy by expanding a heated and partially ionised electrically conducting working fluid, through an intense transverse magnetic field. This causes an electric field within the fluid which can supply power to an external load.

The world's first, significant MHD plant, with a capacity of 200 MW has been on stream for over three years in the USSR and a 500 MW is under construction, to be completed in 1985.

As well as their high conversion efficiency MHD generators have the added advantage of giving off exhaust gases at very high temperatures which could be used either as process heat or to raise steam and generate electricity in the conventional manner using a turbo-generator.

Research into MHD is also underway in the USA and pilot facilities are under construction in Poland, Japan, India and China.

MHD operates most efficiently when coal fired although other fuels may also be used.

Part Three
THE USERS

Chart 19 USA ENERGY CONSUMPTION

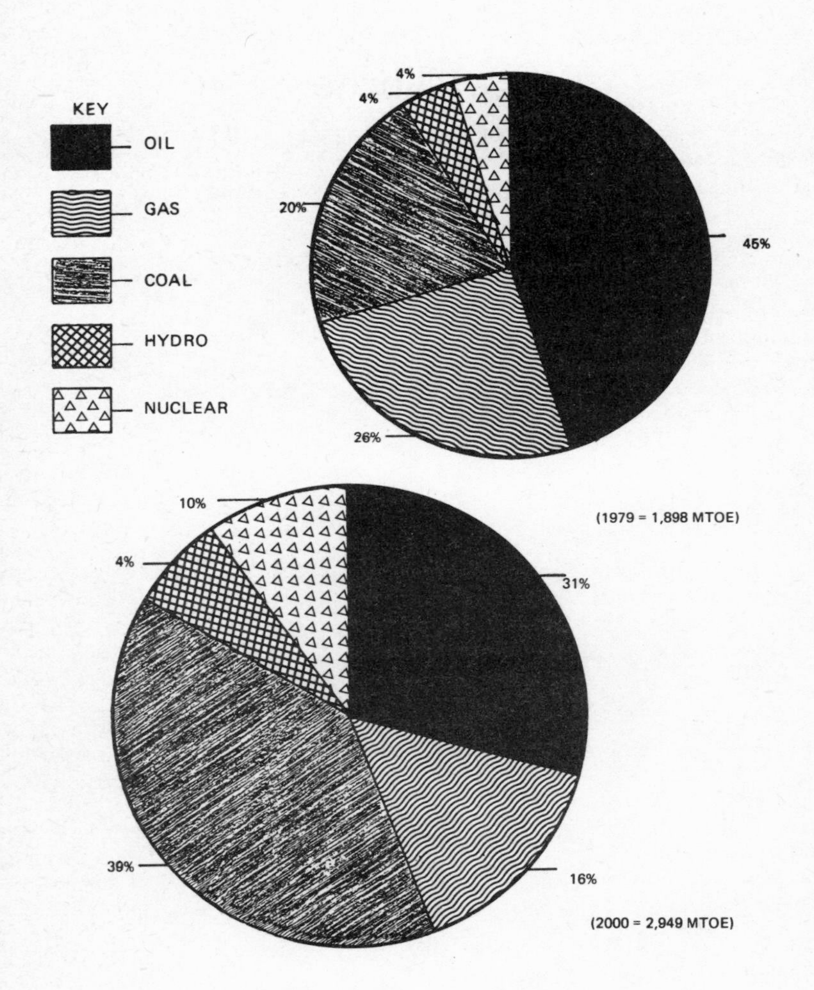

Chapter 1

The United States

The United States is the world's biggest energy consumer. In 1979, out of a world energy consumption of 6,960 million tonnes of oil equivalent, the US accounted for 27%, five times the consumption of Japan, thirteen times that of the continent of Africa, eighteen times that of the Middle East.

The main fuel used at present in the United States is oil which accounts for 45% of energy consumption. The US is at present engaged in a major effort to reduce dependence on oil, particularly imported oil, by the development of alternative fuels and by conservation measures aimed at reducing the energy intensity of the US economy. This aim to make the transition to a more broadly based (more indigenous) supply pattern, is shared by all other industrialised countries but it would be correct to regard the US as being in the vanguard of the transition. As the world's richest country with the technology available and generally speaking, public and government support for the appropriate measures (with the exception of nuclear) the situation is one of 'what the USA does today the rest of the industrialised world does tomorrow' provided, of course, that the resources are available.

The United States is particularly fortunate in its energy resources. For many years it provided a large part of the world's oil from its own abundant reserves. Despite the running down of these reserves it was not until 1975 that the USSR took over from US as the world's main oil producer although the US had become a net oil importer some time before, in the 1960's.

The situation today has changed to where almost half of the USA's huge oil consumption is imported. However, the growing expense, the 'humiliating' dependence and the frequent supply scares have hardened resolve in the US and the results are already becoming apparent. Between 1978 and 1979 oil consumption fell from accounting for 47% to accounting for 45% of total energy consumption. In the future, bigger reductions will be registered with a 15% reduction in actual consumption and a 12% reduction in imports, so that, by the year 2000, American dependence on oil imports will have fallen from 22.1% of total energy consumption in 1979 to 12.5% in 2000. A very significant saving will come from the decline in demand for motor fuel of some 25% by 1990*. This will be achieved mainly as a result of increased fuel

* Bankers Trust Company, US Energy & Capital. A Forecast 1980-90.

efficiency by means of standards implemented by federal legislation. Also fewer passenger miles will be travelled: this will happen because of the increase in gasoline prices to reflect world oil prices (rather than cost of indigenously produced oil). A further consequence of higher gasoline prices and a further contributor to reduced gasoline consumption will be a contraction in the market for passenger cars.

Gas, although reserves are still fairly large, is also running down, having declined from 6.7 billion tonnes oil equivalent in 1970 to 4.7 BTOE in 1980.

However, the US is also the 'Saudi Arabia of coal' with proved reserves of 75 billion tonnes oil equivalent. US consumption of coal is expected to almost treble by the year 2000 to account for almost 40% of total energy needs, with a further 110 MTOE of coal for export. Coal will increasingly replace oil in all areas where fuel is used to produce heat: in industry, the home and in electricity generation. The US economy is likely to grow at between 2½%-3% per annum between 1979 and 2000 (on average). This level of growth will require an energy consumption growth of about 1.8% per year to 1990 and 2.2% per year from 1990 to 2000.

This comparatively high level of economic growth will be made possible, partly by the USA's growing success in lessening dependence on oil imports, partly because of the growth produced by the revitalisation and swift expansion of a new coal industry and infrastructure and partly due to the new synthetics industry which will be making transport fuels and other oil substitute products from oil shales and coal (see Synthetics Section).

In recent years the energy intensity of the USA's economic activities has fallen due to conservation, greater efficiency and change of economic mix away from heavy manufacturing industry and towards more service based industries. The energy intensity of synthetic fuel production and of the coal industry will work against this trend - the reason for the higher energy demand growth 1990-2000. By 2000, synthetics production is likely to be supplying some 6% of US energy demand. Increased production of synthetic oils above this level could lower the need for imported crude oil by a further 90 MBOE or so.

Nuclear power will increase its contribution to supply to about 10% by the year 2000. Although nuclear is the fastest growing of the USA's four primary fuels this is a lower level of growth than was confidently expected two years ago and could, finally, be even lower.

The degree of public opposition which the quadrupling of nuclear capacity over the next 20 years will arouse, is one of the imponderables in any prediction of the US energy situation in the year 2000.

Similarly, the growth of the coal industry could be slowed by restrictive regulations relating to mining operations and air pollution control.

Chart 20 JAPAN ENERGY CONSUMPTION

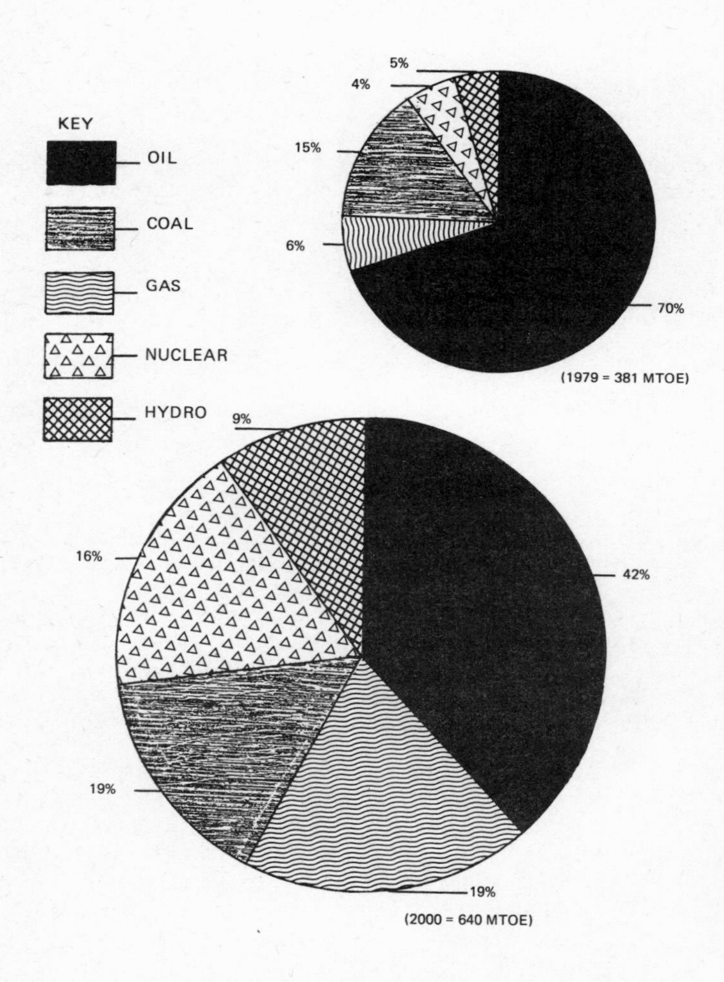

Chapter 2

Japan

It is heavily ironic that the only country to have suffered the effects of atomic warfare should also be the country which would benefit most from nuclear power. Japan has virtually no natural energy resources and relies on imported fuels for 90% of its supply. It also has the dubious distinction of being most heavily reliant on the most vulnerable of all energy sources: 75% of Japan's energy needs were supplied by oil in 1979.

Japan is, of course, alive to its vulnerability and is doing what can be done to reduce oil dependence by using it more efficiently and by developing alternative sources. The Government has set a target for reducing imported oil dependence to 30% by 1990. However, such a reduction would require an extremely rapid increase in nuclear capacity to some 50 gigawatts (GW) from a present level of 15 gigawatts.

Although environmental problems will, as has already been the case, hinder development of the Japanese nuclear industry, by 2000 Japan will have the largest nuclear generation capacity outside the USA at around 80 giga watts and will probably be among the world leaders in the export of nuclear plant.

However, even this swift growth will mean that only 16% of Japan's needs will be met by nuclear power in 2000 and the fact that oil dependence will have been lowered as a percentage of total consumption from 70% to 41% will be as a result of large increases in coal and gas imports. Japan's LNG imports will be particularly large and Japan will be the only major industrialised nation to make extensive use of LNG in power stations.

In 2000 Japan will still be vulnerable to energy supply disruption. Oil will still be the major component (at 40%) of energy supply. Japan obtains 75% of its oil supplies from the Middle East (and in 1979 almost 20% from Iraq and Iran) and in addition has very little participation in international oil exploration (through Japanese oil companies). All this suggests troubled times ahead for Japanese energy supply.

In addition, it should not be assumed that the international LNG and coal trades, will be without their own periodical supply disruptions - the number of cancelled contracts between Algeria and LNG importing countries - should be taken as an omen of future problems here.

As a country with so little natural resources Japan is at a great disadvantage. However, Japan has advantages also, the financial resources that past economic miracles have produced mean that Japan can buy in a more and more expensive market for many years without fear of penury. Also, the national commitment and technological flexibility which has made Japan account for about one tenth of the world's GDP and which have already enabled the country to make huge strides with energy conservation in industry and in the development of nuclear power (such that a commercially viable FBR is likely to appear in the mid-1990's) should enable Japan to remain relatively immune to energy based recession.

But, having said this, it should be remembered that any country with Japan's degree of reliance on OPEC oil should be regarded as delicate despite the appearance of rude good health (with GNP growth 1979-2000 of about 3.7%)*. If OPEC catches a serious chill Japan will be at the front of the queue for pneumonia.

* Growth in energy consumption about 2.5% per annum 1979-2000.

Chapter 3

Western Europe

Western Europe,in 1979 the second largest energy consuming unit after the United States, will, of necessity, experience considerable change in energy mix (the combination of fuels which supply energy demand) over the next 20 years. During that period dependence on oil will be reduced from 54% in 1979 to 38% in 2000, an actual reduction of some 20 MTOE.

This reduction, one of the prime aims of European governments, will be achieved (and can only be achieved) by the committed development of alternative fuels - nuclear power, coal and gas will all expand their contributions - and by conservation (both using less and being more efficient). The transition will be aided by an inevitable movement, common to all the industrialised nations, toward less energy intensive, more service based industries.

The lowering in energy intensity will mean that, despite a growth in Europe's energy consumption between 1979 and 2000 of 1.6% per annum, a growth in GDP of some 2.5% to 2.8% per annum should be achieved, an improvement in energy coefficient over the present level.

The most dramatic increase comes in nuclear power, which, despite the problems of public opinion in a number of countries achieves a sevenfold increase to account for 16% of Europe's energy needs in the year 2000, a larger proportion than natural gas.

By the year 2000 Europe's recent hydrocarbon bonanzas, North Sea gas and North Sea oil, will be on the downward path, both probably reaching their peaks of production in the late 1980's (but with oil production holding its peak for some time). However, despite this, oil imports will still diminish between 1990 and 2000. This will be made possible partly by the resurrection of the European coal industry, which will increase production by some 50% in 1979-2000, but more significantly because of huge increases in gas and coal imports. Imported gas and coal will make up 17% of European consumption in 2000 (as against only 6% in 1979).

The main change in usage making the substitution of oil possible will come in electricity generation, where coal will be used instead of oil wherever possible. Such change, however, is very expensive, it generally being scarcely any cheaper to adapt an oil fired power station to burn coal than to build a completely new station.

Gas will replace oil in industrial, domestic and other uses, as a heat producer. Gas production from the Norwegian sector of the North Sea will not be enough to make up for the decline from the mid 1980's onwards, of supplies from the Netherland's Groningen field and imports from the USSR (perhaps by the proposed new pipelines from Siberia by the early 1990's). Algeria and the Middle East will supplement supply.

The net effect of a great deal of capital expenditure and rethinking in the areas of energy supply will be the desired considerable reduction in oil imports, however oil will still remain in 2000 the main fuel at 38%, and the fact that oil imports still represetn 27% of energy consumption renders Europe still vulnerable to Middle East supply disruption.

A further problem is that imported fuels will still represent 45% of total energy consumption in the year 2000.

Of the five major European consumers, Spain the least industrialised will experience the highest level of growth in energy use. Italy will also experience considerable growth and worsening economic situation due to the difficulty in substituting other fuels for oil.

West Germany, France and the U.K. will all be better able to cut down oil use by increased commitment to coal and nuclear power. France will probably overtake U.K. as the second largest energy consumer (toGermany) in Europe with nuclear power making up a unique 34% of energy consumption.

Chapter 4

United Kingdom

The U.K. has the largest hydrocarbon reserves of any country in Europe and in 1980 achieved a net self-sufficiency in oil. However, the size of the oil and gas reserves - 2.4% and 1% of world reserves respectively - provide no more than a useful but fairly short lived breathing space as the world's fossil fuel resources diminish. By the early 1990's oil production from the U.K. sector of the North Sea will reach a plateau. Gas will peak then begin to decline before this, therefore it is just as much the U.K.'s vital concern to reduce dependence on oil as it is for her less well endowed European neighbours.

Between 1974 and 1979 the U.K.'s low level of economic growth meant also a low level of energy demand increase of only 0.6%. This level of energy consumption growth will be, on average, maintained to the year 2000, though this does not mean that gross domestic product growth will also parallel the 1974-79 level of increase. The achievement of a modest, but improved average yearly level of economic growth to the year 2000 will be possible as a result of the reduction in energy intensity of the U.K. economy by dint of conservation and a continued shift in industrial emphasis towards lighter and more service based industries.

Oil demand will decrease mainly as a result of the reduction in its use in power stations and elsewhere as a heating fuel. Price will be the main mechanism by which this will be achieved. The gap left by oil will be filled by increased consumption of coal and an expansion of nuclear generating capacity. By the year 2000, natural gas consumption will be at about the same level as in 1979 due to the run down of supplies from the North Sea. During the 1990's both imports of LNG and the manufacture of SNG (substitute natural gas) from coal will become more significant.

However, although oil use in the U.K. will decline, consumption of petrol and DERV (diesel engine road vehicle), which accounted for about 30% of petroleum product deliveries in 1979, will increase. The car population will be 1.5 million higher in the year 2000 than in 1979 at 22.5 million and road transport fuel consumption will rise by up to 20% above the 1979 level.

It is ironic that the era of U.K. indigenous oil is also the time when coal, the fuel of the industrial revolution, is experiencing a resurgence . The

Chart 21 UNITED KINGDOM ENERGY CONSUMPTION

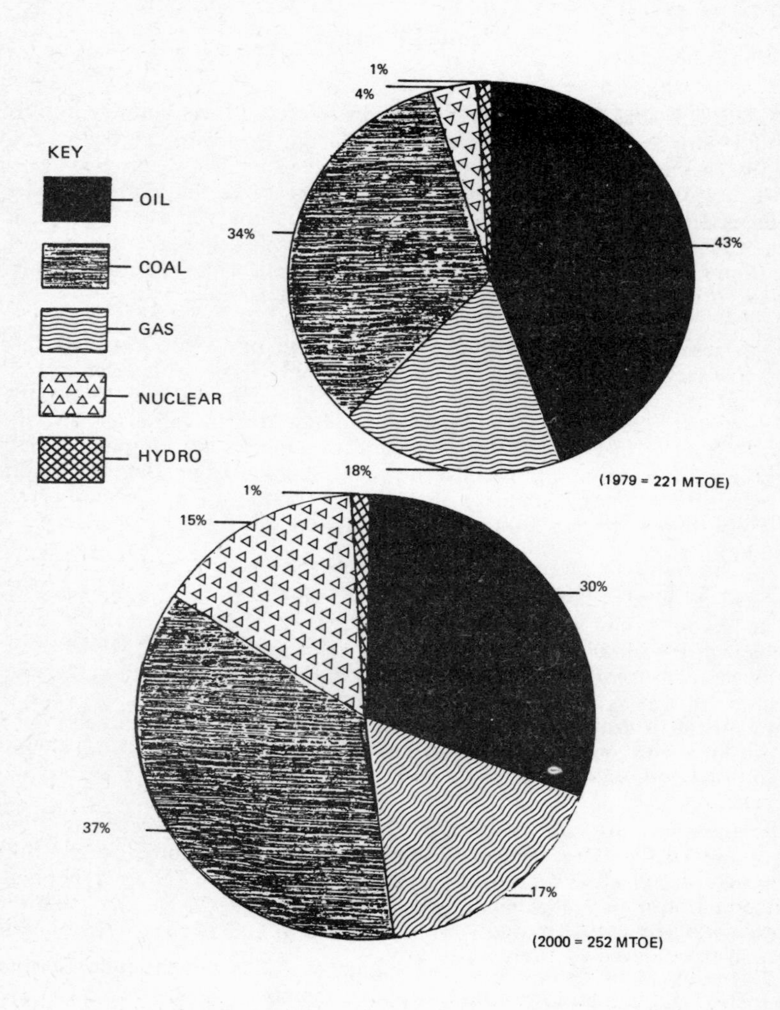

The U.K.'s economically recoverable coal reserves at present are the equivalent of some 2.68 billion tonnes of oil, almost equal to the sum of proven North Sea oil and gas reserves (2.709 BTOE). However, the U.K.'s geological reserves of coal are far higher (127 BTOE) and progressively more of these reserves will become economical as world energy prices rise.

At the beginning of the 1970's the U.K. had the largest coal industry in Western Europe, although even at this time, a rapid decline was in progress. The fourfold oil price increase of 1973-4 caused the adoption in 1974 by the U.K. Government, the National Coal Board and the National Union of Mineworkers of the 'Plan for Coal' which set production targets of 170 million tonnes of coal per year (114 MTOE) by the year 2000. Present production stands at some 120 million tonnes (80 MTOE).

The level of production proposed in the plan for coal will not be achieved. Nor, were it to be achieved, would the demand be there. Demand for coal, will by the year 2000, have increased from 34% in 1979 to 37% of total energy consumption and possible environmental problems over the siting of new mines may mean that, even to meet this level of demand, imports will be necessary. A factor which could, in the short term, harm the coal industry is that imported coal, even from as far away as South Africa, the USA or Australia, is cheaper than U.K. coal due to lower mining costs. Subsidies or other legislative measures will therefore be necessary in the short term if the U.K. electricity industry is to continue to use, almost exclusively, British coal in its power stations.

Ultimately however, the U.K. coal industry will achieve its expansion to production levels of 170 million tonnes per year and above, by virtue of rises in world coal prices and increased coal demand in the U.K. for electricity generators and for the manufacture of synthetic oil and synthetic natural gas. However, this will not be until well into the 21st century.

From being the pioneer of commercial, nuclear produced electricity, the U.K. has fallen well behind in the nuclear stakes. In 1979 nuclear power accounted for only 4% of total U.K. energy demand and 11% of electricity demand.

Although factors such as low electricity demand, public opposition to nuclear power (which could prove particularly effective against the introduction of a PWR at Sizewell) and the poor record of the British nuclear industry on its construction sites, an increase in capacity to some 25,000 MW should be achieved by the year 2000.

The level of increase in U.K. energy consumption envisaged by this study is below the lower estimate in the Department of Energy's paper, 'Energy Projections, 1979'. The Department of Energy's low estimate was based on an assumption of economic growth of about 2% P.A. to the year 2000. This seems unlikely now, mainly because the world crude oil price rises in 1979 and 1980 mean that the Department's assumption that world crude prices would rise to $30 a barrel (1977 prices) by the year 2000 will turn out to be a significant under estimation.

Usage trends now establishing themselves in the U.K. show the likely pattern of change for the rest of the century. The shift in industrial emphasis will mean less fuel used in the iron and steel and other heavy industrial sectors and more in less energy intensive industries.

Energy use within the domestic sector will grow very slowly and will make up a progressively lower proportion of total energy use. Conservation, encouraged by higher prices, will be a major cause. Energy saving will be effected both by changes in energy usage habits and by increasing attention to insulation standards in new houses.

Transport will be a growth area: by the year 2000 this sector will be accounting for almost one quarter of all energy used in the U.K. Road transport will account for some 78% of this (air transport 13% and rail and marine transport 9%).

In 1980 (January-October) 78% of all electricity generated in the U.K. was produced in coal fired power stations, coal's biggest share for the past five years. Oil fired power stations were responsible for only about 10% of electricity generation - 5% less than in 1979 and although this was partly due to the fall in electricity consumption the trend away from oil will continue. Growth in electricity consumption between 1979 and 2000 will be fuelled by the expansion in coal and nuclear power generation.

Electricity consumption is likely to increase by 1.2%-1.6% per annum, less than the rate predicted in the Department of Energy's 'Energy Projections 1979', but greater than the rate by which total energy consumption will increase. By the year 2000 nuclear generated electricity will account for 33% of the electricity generated in the U.K. with coal the major fuel at 60% and oil only supplying 5% of generation.

In the Department of Energy's 'Energy Projections 1979' it was predicted that following a period of energy self-sufficiency in the 1980's, the 1990's would see a return to fuel imports ranging between 4%-11% of volume consumption in 1990 and 8%-23% of volume consumption in 2000. Economic and energy consumption trends subsequent to the Departemnt of Energy's forecast suggest that a surplus of indigenous production over consumption may be maintained for the remainder of the century, in value and, probably in volume terms. This should prove of great benefit to the U.K.'s balance of payments.

Maintenance of energy self-sufficiency is dependent on four conditions: low growth in energy consumption caused by modest economic growth and increased conservation, the fourfold increase in the U.K.'s nuclear capacity, the maintenance of increased coal production and the acceleration of the exploration and development of the United Kingdom Continental Shelf's oil reserves as oil becomes less easy to find.

The expansion of the coal and nuclear industries requires a high level of public consent and political will, beset as both industries are by environmental problems. A higher degree of Government support for and identification with the nuclear industry and a systematic spelling out of the simple arithmetic of energy supply and demand and their bearing on standard of living will be necessary in the future, as environmental lobbies become better organised.

The oil industry, on the other hand, is not suffering from too little Government involvement but from too much. If the huge sums of money needed to finance exploration for oil where the finds are smaller and the depths are greater, are to be invested, then the fiscal reins on the oil companies should not be tightened with such frequency. When tax increases on North Sea oil make marginal discoveries uneconomical to develop or inhibit exploration, the increased tax benefits to the exchequer are illusory since they bring closer the end of oil self-sufficiency and the reestablishment of crude oil imports as the big discoveries - Forties, Brent Ninian etc. - run down.

Chart 22 WEST GERMANY ENERGY CONSUMPTION

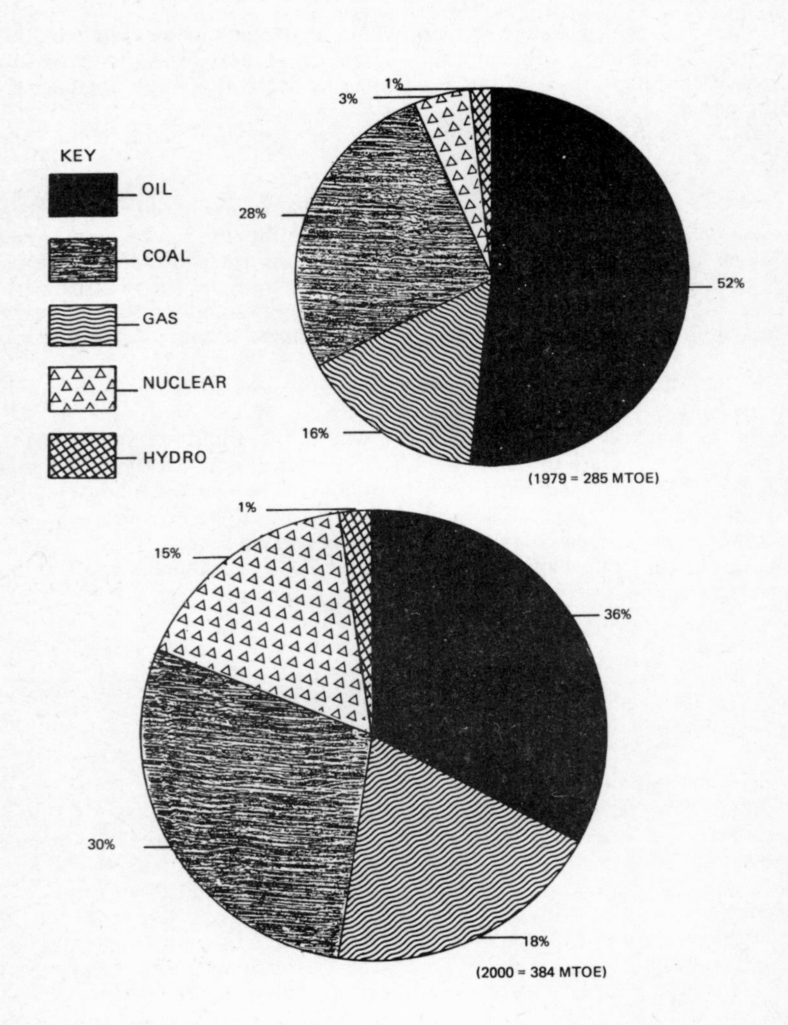

Chapter 5

West Germany

West Germany is Western Europe's energy heavyweight accounting for more than 20% of energy consumption. West Germany also has the dubious distinction of being Europe's biggest oil consumer and importer although France, Italy and Spain are all more dependent on crude oil at 60%, 68% and 61% respectively of total consumption as against West Germany's 51.5%.

The reason why West Germany only relies on oil for just over half its needs (there is some domestic production but this has little significance at 4.8 MT in 1979) resides in the country's considerable coal deposits - over 5 BTOE of economically recoverable reserves - and considerable natural gas imports. Nuclear power's contribution is minor.

The problem with Germany's coal is that, like the U.K.'s, it is very expensive to mine and can only remain competitive with imported coal (and even with imported fuel oil and gas) through huge government subsidies (6.3 billion DM in 1979).

Despite the high production costs, however, a massive investment programme is underway to raise coal production to some 100 MTOE by 2000, which will require capital expenditure of some 8 billion DM for new mines and 3 billion DM to maintain the old ones.

But, West Germany's vulnerability to oil supply disruption means that the cost must be borne, because in the high cost energy future, not to exploit any indigenous fuel resource is madness.

West Germany has all the usual incentives of the large crude oil importer to reduce oil dependence. Firstly, fear of supply disruption; there has already been one major scare. Before the Iranian revolution, Iran was West Germany's largest supplier. The dramatic reduction of oil from this source was, in the event, met by increases in supply from Saudi Arabia, Libya, Nigeria and the North Sea.

Secondly, rising world crude prices have meant that paying for oil imports caused in 1979 a huge deficit in the Federal Republic's current account balance - the first since 1965. Oil price rises since 1979 will mean that the deficit will rise to about DM30 billion in 1980. But despite these strong incentives and a government policy called 'Away from Oil', little progress in real oil usage reductions has been made. 1979 oil consumption was the highest ever at 147 million tonnes and although consumption will certainly be lower in 1980, this will mainly be as a result of recession and warmer weather although some of the reduction is also attributable to some substitution for oil, coal and gas based processes in industry.

In the early 1950's oil was responsible for under 5% of West Germany's energy consumption; gas was scarcely used at all and coal was the dominant fuel. Oil became the main energy supplier during the 60's with the coal industry sinking deeper and deeper into the doldrums and coal consumption reaching its nadir in 1977. Production in 1979 was the highest for four years but coal still accounted for only 28% of energy consumption. With oil, gas consumption has grown significantly during the seventies from accounting for under 5% of energy needs in 1969 to 16% in 1979. 39% of West Germany's gas consumption was indigenously produced but unlike coal reserves, West Germany's gas reserves are limited. 1980's reserves were almost 40% down on the 1970's and under 10 years worth of production at today's level is possible. The remainder comes mainly from the Netherlands, a source which will also begin to decline soon, and from the USSR (16%). West Germany could, undoubtedly, maintain or even increase its consumption of natural gas should the 3,000 mile gas pipeline from Western Siberia be built. However, the cost would be great, and the project, once agreed, would take around five years to complete and this expansion of gas consumption will mean a speedy increase in the level of energy imports, which would have to accelerate in the 1990's as a result of declining domestic production.

There was a time around five years ago when West Germany seemed about to become Europe's leading nuclear nation. However, since 1975 the West German nuclear industry has expanded little. In 1975 nuclear power supplied 2.1% of energy demand. In 1979 that had increased to only 2.8%. The main reason has been that in West Germany opposition groups have found it relatively easy to prevent construction of nuclear power stations through court injunction. They have been able to do this on the grounds that insufficient provision has been made for the disposal of spent nuclear fuel - a contention made possible by the refusal of the State of Lower Saxony to permit the construction of an integrated nuclear storage and reprocessing plant at Gorleban.

However, ultimately, the necessity for a significant expansion of nuclear capacity will provide the impetus to resurrect the West Germany nuclear industry and by 2000, West Germany should increase the contribution to energy supply made by nuclear power, to 15%. Unlike the UK, West Germany does not have technical problems with nuclear stations, the West German nuclear industry has taken the American PWR and, by redesigning certain features, has (it is claimed) raised the standard of performance, reliability and safety. This was substantiated in a study by Imperial College, London which examined the performance of reactors 1965-78 and discovered that of 106 light water reactors in operation, the ones built in Germany over the years 1968-78 were the best performers (see also Nuclear Power Section for 'reactor performance').

But even though a substantial expansion in nuclear power will ultimately take place it is to coal that West Germany is looking primarily, to reduce oil dependence. Coal consumption should rise by 47% between 1979 and 2000. As well as making an increasing contribution to electricity generation, coal should be beginning to replace oil in some of its, hitherto unique, areas of usage. The Government is committed to promoting the development of coal liquefaction and gasification technology and feasibility studies for West Germany's first commercial scale coal liquefaction plants should be completed early in 1981.

But it is unlikely that synthetic gas or coal will contribute more than 3% of oil and gas demand by 2000.

In the years to 2000 West Germany will suffer financially i.e. by having to pay out for imported fuels, for two reasons. Firstly, because of the long hiatus suffered by the nuclear industry. Secondly, because West Germany was slow in expanding its coal industry, not heeding the warning of the oil crisis of 1973 and starting late. West Germany will probably not be able to expand coal production fast enough to meet its needs and the mountain of DMs already being shelled out for crude oil - West Germany will still be Western Europe's biggest oil importer in 2000 - will be made a little higher by net coal imports which will probably be some 16 MTOE.

However, despite being slow off the mark in effecting its transition away from oil, the large increase in energy consumption over the next 20 years (35%) unlike Italy's, is a bullish indication that the West German industrial juggernaut will be rolling as strongly as ever.

Chapter 6

France

Despite its paucity of hydrocarbon reserves, France is on a course which will make it the major European consumer least dependent on oil by the year 2000. The course is nuclear and by the end of the century a greater proportion of France's energy will be supplied by nuclear power than any other country in the world. Why has France been able to avoid the nuclear malaise that has sapped the world's confidence in the atomic solution since Three Mile Island. There are a number of reasons.

Firstly, France, like Japan, has little alternative but to go heavily nuclear with only small reserves of expensive-to-mine coal available, and one of the highest levels of dependence on oil in the whole of Europe (62% in 1979) Thus the choice for France was a simple one - to make one's own energy or buy it at high and increasing prices. It was in this simple way that the nuclear issue was presented to the French people. The French Prime Minister Raymond Barre warned in 1979 that France must choose between nuclear power and a lower standard of living. But it would be misleading to imply that the matter of nuclear power was ever presented as a matter of choice to the French. A cabinet minister was reported as making the comment 'We don't consult the frogs before we drain the marsh'.

This mischievous remark goes far in isolating the difference between France and the rest. It is in countries where control is heavily decentralised - where the devolution of executive power makes regions able to withdraw cooperation or the laws of the land allow pressure groups to prevent construction of power stations by court injunctions, that nuclear programmes have been slowed and stopped. The French approach is to go ahead and build nuclear power stations regardless of objections and so far this has worked.

At present the French programme, which aims at 50,000 MW of nuclear power by 1990 is only months behind schedule and although it is unlikely that the programme will be achieved without some setbacks - the issue of hairline cracks in essential metal components of a number of pressurised water reactors may be such a setback - the tide at present would seem to be moving inexorably toward the desired nuclear future.

The fact that the majority of the public favours nuclear power in France is also due to the heavy advertising and numerous public forums which have brought home the significance of the energy crisis to the people. The Government has stated that nuclear power is the only answer to the energy problem and has put forward four strong reasons why. Firstly, the switch to nuclear power will help the balance of payments by substituting indigenous uranium for imported oil. Secondly, the emphasis on energy supply will be shifted away from the unstable oil exporting Middle East. Thirdly, the cost of electrical energy will be reduced because nuclear produced kilowatts are cheaper than those produced by coal or oil (electricity from PWR was said to cost 8.68 centimes per KWH (kilowatt hour) as against 12.68 for coal produced electricity and 13.33 for oil produced electricity). Finally the massive expansion of French nuclear industry gives France scope to become a major nuclear plant exporter.

Of course, so huge a commitment to construction of expensive and sophisticated power stations is taking up a huge slice of France's investment finance but with 70% of energy requirements at present imported it is difficult to think of a more prudent investment.

In 1979 France's energy bill came to about FFr78 billion, of which about FFr65 billion was for imported oil. By 2000 oil prices will be higher but with oil imports almost halved and oil accounting for only 25% of total energy as opposed to 60% in 1979 with 34% of energy supplied by indigenously produced nuclear energy, France should be in a strong position.

The success of France's energy transition is the reason why the French economy should achieve a good level of growth to the year 2000, its buoyancy expressed in the fact that between 1979 and 1990 France will become Europe's second largest energy consumer after West Germany displacing the much more slowly growing U.K. to third place.

In other words, the fact that France's total energy consumption will increase by some 35% (the same rate as West Germany) over the next 20 years, 2½ times the rate in the U.K. is not an indication of an increasing or even a static level of energy intensiveness within the French economy. Although the cornerstone of French energy policy is nuclear power, considerable investment is also intended in alternative energy sources, substitution of fuels and conservation.

The way in which the dramatic oil savings will be achieved will be by firstly a dramatic cut in oil heating in industry where two out of three power plants will be changed and coal burn will be increased from 2.7 to 16 million tonnes (i.e. 1.8 to 10.7 MT Oil Equivalent). Second, a 20-30% increase in the standard of insulation in new houses and offices plus improvements to the insulating of many existing buildings. Thirdly, increasing the use of renewable energy sources such as wood, solar and geothermal power which are intended to supply up to 10% of energy needs by 1999 (hydro electric power, and geothermal power which is used in a number of district heating schemes, in 1979 supplied 14.5% of energy demand). Fourthly, by means of a 30% reduction in the petrol consumption of new vehicles.

Investment in these measures is intended to be in the region of FFr70-90 million per year.

Chapter 7

Italy

Italy has been living on a knife edge for some time to avoid the severe consequences of not being able to get enough fuel to meet energy demand. Italy is the most dependent on imported crude oil of the five main European energy consumers - 70% of the country's energy needs are supplied by oil. In 1980 Italy was faced with the prospect of an oil shortfall of 25 million tonnes or a quarter of oil supplies. This economically disasterous prospect was, at least, mitigated by the decision to ease price control on oil products, thus enabling the country to compete for crude on the world market. Although the crude oil situation eased later in 1980, Italy has been formulating plans to prevent future energy supply scares.

After oil, gas was Italy's second most important fuel in 1979, accounting for 15% of total energy needs. Italy has a medium resource of natural gas, producing, in 1979, some 12 MTOE indigenously - roughly half of total consumption.

Hydro electric and geothermal schemes accounted for 8%, coal,(of which Italy has very little)for 7% and nuclear power for only 1%.

The fact that Italy's oil import bill in 1980 will be around $17 billion coupled with the fact that because of its excessive dependence on oil Italy has been lurching from supply scare to supply scare ever since 1973 has now seemingly convinced the Government that long term planning rather than ad hoc solutions must be sought.

The prime need is to reduce oil dependence but the risk of inadequate electricity generation capacity and the spectre of 'brown outs' and blackouts are also very much a current concern. As a country with very little indigenous fuel reserves, Italy would seem tailor made for the French solution - massive nuclear expansion - and a ten year plan has been put forward for ENEL (Ente Nazionale per l'Energia Elettrica) to invest about $25.5 over the first five years of the plan in the construction of five twin reactor nuclear power stations with an overall capacity of 10,000 MW. In addition coal fired stations to a capacity of 13,500 and hydroelectric plants of 1,800 MW, turbogas stations of 1,130 MW and geothermal plants of 100 MW are proposed all of which, if implemented, is intended to increase Italy's electricity generating capacity to 75,000 MW by 1990.

The purpose is that by 1990 oil should only contribute some 42% of the fuel for electricity generation.

One problem with the plan is the seeming inability of Italy to achieve a real commitment to nuclear electricity. Having discussed nuclear expansion for years very little has been achieved; however the pressure of necessity should, during the years to 2000 enable a fairly substantial nuclear programme to be implemented such that around 11% of Italy's energy needs should be supplied by nuclear power in 2000. The increase in coal imports to make possible a degree of switchover from oil to coal generation should also be achieved with coal accounting for 12% of total energy in the year 2000. All but around 2 MTOE of Italy's coal in 2000 will be imported but this will be a far cheaper and more easily available fuel by then.

Apart from nuclear power the main expansion in fuel use will be in natural gas. Consumption should almost double over the next 20 years. The new supplies will come mainly from the 2,500 kilometre 'Trans-Med' pipeline which is now under construction between Algeria and Italy. Imports of gas through the pipeline are expected to reach 12.36 billion cubic metres by 1984 (10.6 MTOE).

Even with the implementation of these measures it is difficult to see oil falling below 1979 levels before the year 2000 and Italy will still be vulnerable to world oil supply as well as suffering the financial stresses of a huge import bill for imported fuel and the capital costs of the nuclear programme and the switch away from oil in the electricity sector.

Italy's fairly high level of energy growth will not represent a correspondingly high rate of economic growth and will lag behind countries like Germany and France in the implementation of effective conservation measures.

Chapter 8

Spain

With little in the way of indigenous fossil fuel reserves apart from some low quality coal and an import bill for oil in 1980 which will be around $12 billion - double the previous year, Spain is another country where the need to reduce oil dependence as quickly as possible is paramount.

How to do it, however, has been a question to which the answer has changed over the past two years.

In 1978 a National Energy Plan (NEP) for Spain was agreed on, of which the main element was a speedy expansion of the nuclear power programme. In 1979 nuclear power only accounted for 2-3% of total demand while oil reached its highest level since 1976 of 65.5% of total demand. The plan expected 10,500 MW of installed nuclear capacity by 1987 thus reducing dependence on oil fired plants as well as expanding the economy and providing jobs.

A year later the NEP was changed. Although the aim of reducing oil dependence remained, much greater emphasis was laid on coal as the most effective means of doing so rather than nuclear power.

This change of heart was partly yet another legacy of the severe blow dealt to the credibility of nuclear power by the public relations disaster of the Three Mile Island PWR failure.

Resultant uncertainty has had the effect, in Spain, as elsewhere, of forcing the Government to take a firmer line on safety regulations and procedures. Secondly, a strong anti-nuclear lobby has grown up in Spain which has not only affected the NEP but the present nuclear power station construction programme. The Plan now proposes that instead of 14.7% of primary energy being supplied by nuclear power in 1985 the proportion should be 11.9%.

However, in the present political climate, with, for example, the militant Basque separatist movement ETA threatening to prevent the Lemoniz (near Bilbao) plant ever operating, even this level of nuclear growth may be optimistic. However, if oil use is going to be replaced at all a significant proportion of nuclear produced energy will certainly be necessary by 2000 and levels of 17% of total energy needs by 1990 and 25% by 2000 should be possible.

The question of nuclear electricity is complicated by Spain's large hydroelectric contribution to the grid, which according to the level of rainfall can contribute between 6% and 12% of total energy needs. Since nuclear power stations are only efficient when operating at full load, the large variation in Spain's electricity supply needs a certain proportion of oil or coal fired plants to provide a margin of flexibility.

Coal is therefore a major part of the plan. Spain has large coal reserves (though a large proportion is of low quality) but in addition coal is cheap to import. Also coal fired power stations are of known technology and cheaper and quicker to build than nuclear stations as well as lacking the public acceptance problems.

Thus coal should increase its contribution to Spain's energy consumption from 21% in 1979 to 25% in 1990 and 29% in 2000.

This will mean some 15 million tonnes of coal imports by 1986 and more later and the Government will be granting credit facilities to industries which undertake to convert from oil to coal.

These measures should enable a reduction in Spain's oil dependence from 63.5% to 42% in 1990 and 39% in 2000 but, certainly during the years to 1990 there will be little reduction in actual oil consumption (from 47 to 51 million tonnes a year, The amount will vary according to the needs of the electricity industry i.e. the speed at which coal burning plants may be brought on stream and how much hydroelectric power may be produced in any given year.

So, even by 2000, Spain will still be heavily dependent on oil (44% of total), and meeting oil needs will continue to be a problem in the short term since Spain, relied on Iran and Iraq for 20% of its oil before the Iran-Iraq war and though insulated from shortfall by high levels of oil stocks at present will need to find alternative supplies elsewhere should the loss of production in Iraq and Iran continue.

Spain's gas consumption is likely to be fairly insignificant in 2000 with some increase as a result of purchases made from the proposed Siberia/Western Europe 'Yamal' pipeline. However, there is also the possibility of a further trans-Mediterranean pipeline from Algeria to Spain which, if ever constructed, could make a significant addition to Spain's energy options.

Chapter 9

USSR

The USSR has always been a net exporter of energy. Whether this will continue is not only important in a world energy supply context but in a world political context. Competition between the USSR and the Western World for the Middle East's oil is viewed by many as one of the more likely scenarios for a world war. Unfortunately, reaching definite conclusions as to whether or not the USSR can make ends meet for the next 10 or 20 years is hampered by lack of information.

The pessimistic view is as follows - the USSR will soon be unable to meet first COMECON's, then its own oil demand. It will then seek oil on the world market but, lacking the large revenues from exports to which it has grown accustomed will find payment difficult without sacrificing economic growth. Therefore, rather than accepting economic decline, and being in possession of both the military capability and the strategic advantage of Afghanistan, the USSR will take the oil it needs from whichever Middle Eastern country it wishes.

The CIA has been one propagator of the pessimistic view. In 1977 the CIA predicted that Soviet crude production would peak in 1980 at between 11.8 and 11.9 million barrels per day (588-593 million tonnes per year) - production in 1979 was 586 million tonnes. However, due to an inadequate level of exploration, by 1985 that level of production would fall to around 10 MBD (498 million tonnes per year) thus creating the need for a very high level of imports which would be virtually impossible to meet.

A number of more recent occurrences have been taken as support for the CIA view. The Chairman of GOSPLAN, the Soviet planning agency, lowered the official 1980 production target from 637 million tonnes to 602 million tonnes. Also, although the USSR has promised to increase its energy supply to the COMECON bloc by 20% 1981-85 there will be no increase at all in oil supply and COMECON countries have been told that they must meet any shortfall through their own deposits.

In addition there have been a number of criticisms, some from within the USSR, of the Soviet oil industry - such as over exploitation of existing fields by over-use of water flooding and drilling too many production wells and

Chart 23 USSR ENERGY CONSUMPTION

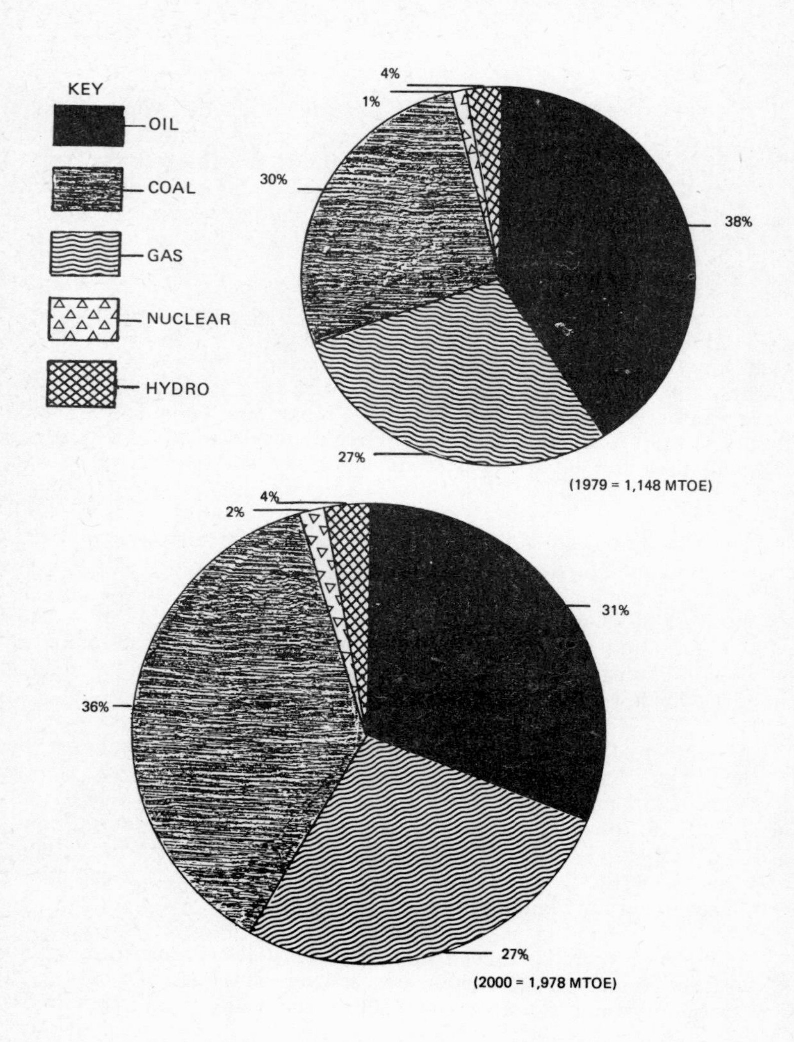

neglect of exploration. One report described this as being, in terms of metres drilled, 30% lower than target. The Russians have also been accused of being 30 years behind their American counterparts in drilling technology, taking a year to drill a 1,000 feet well when an American company could do it in 34 days.

It is certainly true that the USSR has been experiencing a lack of technology in its search for oil and gas, particularly in offshore areas. There have also been reports that during the first half of 1980 Soviet oil production (though ahead of 1979) was, in fact, below the target of 606 million tonnes by some 10 million tonnes and that poor quality equipment was partly to blame.

Looking to the alternative fuels which might make up the gap left by a possible oil shortfall, the following arguments are put forward on the pessimistic side.

Developing nuclear power, which supplies at present only 1% of Soviet energy would take much too long to make up for a large oil shortfall occurring within 5 or 10 years. Coal and gas reserves, though large, are said to be too far from the main centres of Soviet consumption and also suffering from poor technology.

All these arguments, if accepted, could lead one, as it has Saudi Arabia, to believe that the USSR's presence in Afghanistan is part of a grand strategy to capture the Gulf.

But is the USSR really about to become an energy importer? This study considers it highly unlikely.

The USSR is richer in fossil fuels than any country in the world, its proven reserves of oil, gas and coal making up 16% of the world's proven total - 10% of the oil, 22% of the coal and 44% of the natural gas.

However, there is a great deal more oil, gas and coal in the USSR than this. It has been estimated that there is a further 20-35 billion tonnes of undiscovered oil in the USSR - between 14% and 25% of the world's total resources, while 45% of the world's geological resources of coal are said to be in the USSR.

It should also not be forgotten that, despite present difficulties in exploration and in production technology, in 1979 the USSR's total oil exports reached a new high - crude and products exports amounting to about 160 million tonnes.

The Users

It is true that since reaching a high of 10.9 billion tonnes in 1975, USSR's oil reserves have fallen to 9.1 BTOE thus demonstrating that oil discoveries have recently fallen behind production but in view of the huge potential reserves it is difficult to imagine that the USSR will not increase its exploration effort. The incentive, after all, is irresistible. In 1979 the USSR's revenues from oil exports were more than $20 billion. The view that the USSR is determined that both exploration will be stepped up and that obsolete and sufficient recovery techniques will be improved is borne out by reports that the USSR has begun to carry out the most thorough reform ever of its oil industry.

One of the reasons put forward by Oleg Bogomolov, director of the Institute of World Socialist Economics, for the non-increase of USSR oil exports to COMECON was that COMECON was wasteful with its energy consumption with a higher level of energy use per unit of GNP than EEC countries. This in itself would make sense and the promise to increase energy supply as a whole by 20% through increased deliveries of electricity and natural gas could be taken as an incentive for COMECON to make the transition away from oil.

As with the rest of the industrialised countries,the USSR will certainly attempt to reduce its dependence on oil.

However, certainly until the 1990 s there will be no actual reduction in USSR oil use. The USSR's average annual growth in energy use between 1979 and 2000 will be higher than either that of USA or Europe at about 2.6% per annum (as opposed to 1.9% and 1.6% respectively) and in the context of this kind of growth, oil consumption will not fall although the proportion of USSR energy consumption that it represents will decrease from 38% in 1979 to 31% in 2000. At 38% in 1979 however, USSR oil dependence is already lower than in either USA (45%) or Western Europe (54%) while that in COMECON is even lower (23%) therefore it could certainly be said that the USSR is, due to its higher level of development of alternative fuels, starting the period which the future will look back on as the time when oil supply became tight, with a distinct advantage.

Apart from nuclear power, which should experience a fourfold increase to the year 2000 but will still only represent 2% of USSR energy consumption, coal will be the growth area, at least as far as home consumption is concerned, increasing its share of Soviet energy consumption from 30% in 1979 to 36% in 2000 - a huge increase of 370 MTOE (more than Latin America's total energy consumption in 2000).

The USSR has the indigenous resources to follow any path it wishes in reducing the proportion of its energy accounted for by oil; its reserves of gas and coal are both massive while the USSR would certainly not be prevented by public opinion from expanding nuclear power however much its leaders desired. It will therefore follow that path which best combines self sufficiency at the lowest cost with the capability to make the most money out of exports.

Thus nuclear power will be modestly expanded, enough to be sure that the Soviet nuclear industry has the capability, experience and expertise to increase its thermal nuclear capacity whenever necessary while at the same time ground is being well prepared for the next phase in nuclear power - the fast breeder reactor - and the one after that - nuclear fusion.

Thus oil consumption will expand at the slowest rate with increased exploration effort looking to the future and probably the introduction of improved recovery technology aimed at making the best of existing reserves. Exports will probably be continued into the early 1990's picking up added revenue from rational trading on the spot market.

Coal will be the fuel to bear the brunt of Soviet energy increase and replace oil. Gas consumption will increase making the USSR the largest gas consumer in the world.

Coal production is likely to, more or less, keep pace with domestic consumption with exports of some 27 MTOE in 2000 but gas, easier to transport than coal, and commanding a higher price will be the USSR's big moneyspinner as oil imports decrease, especially when, as will probably occur sooner or later, gas prices are linked to crude oil prices.

The problem for the Soviet Union's gas industry, already stretched by having more than doubled production over the past 10 years by projects in some of the world's harshest regions, is not one of resources - huge new gas reserves are still being discovered- but of the expense involved in constructing pipelines thousands of miles long from Siberia where much of the gas is located to the export markets of Europe. This problem now appears to be being solved by getting the countries who want the gas to help build the pipelines - providing investment, expertise and loans against future gas supplies.

In 1979 the USSR supplied around 21 billion cubic metres of natural gas to Western Europe (18 MTOE), 10 billion cubic metres to West Germany (8.6 MTOE) and 11 billion cubic metres (9.5 MTOE) to Italy, Austria and France. A proposed 2,800 mile new pipeline from Siberia's Yamal Peninsula to Western Europe would add a further 40 billion cubic metres per year (34.4 MTOE) of which West Germany would receive around 20-25% with France, Italy, Holland, Belgium and Austria taking large shares and Spain and Sweden also possible buyers. The project would cost some $12 billion and much of the finance is expected to come from West German banks although there have been reports that Japanese financing has also been requested.

The field on the Yamal Peninsula is estimated to contain 20 billion cubic metres of gas (1.72 billion tonnes of oil equivalent) almost 3% of world proven reserves.

In 1979 the Netherlands was the world's largest gas exporter at 38.8 MTOE. By 1990 the USSR will have taken over this position, exporting around 44.3 MTOE but with a gas field such as the Yamal there is plenty of room for increase.

Chapter 10

Eastern Europe (excluding Yugoslavia)

In a time when the world's industrialised nations are striving to develop alternative fuels in order to reduce their potentially debilitating dependence on oil, the nations of Eastern Europe have one great advantage - they are much less dependent on oil for their energy than any of the world's main energy users with the exception of China. Less than one quarter of Eastern Europe's energy consumption in 1979 was accounted for by oil, as compared with over half in Western Europe.

Coal, accounting for over 60% of energy use, is the main fuel in Eastern Europe. Substantial coal reserves exist in Eastern Europe, particularly in Poland which was the world's second largest coal exporter in 1977.

Remaining needs are met by oil 23% and gas 14%. Oil reserves in Eastern Europe are small. Rumania had about 200 million tonnes in 1979, Hungary about half as much. Most of Eastern Europe's oil, around 75%, is supplied by USSR.

Problems for Eastern Europe could however be on the way: Mr. Oleg Bogomolov, director of the Institute of World Socialist Economics, said in 'World Marxist Review' that for the next five years Soviet exports to Eastern Europe (including Yugoslavia) would remain at 80 million tonnes per year. With oil usage already at a very low level, the fact that oil may have to fall as a percentage of total energy consumed, having risen slowly for the past 10 years could cause economic problems should it be necessary to bid for oil on the world market.

The brake on gradually increasing USSR oil exports will be mitigated by a predicted 20% increase in total energy supplied by USSR to COMECON during the next five years but accommodation to a swiftly changing mix of fuels may prove expensive and difficult.

Gas, nuclear power and increased usage of coal, will be the means by which oil's role in Comecon's energy economy will be further reduced over the next 20 years. Modest but useful gas reserves exist in Rumania, Poland, Hungary and East Germany while completion of the Orenbury gas pipeline will also prove helpful in enabling USSR to make good its undertaking to increase energy supplies to Comecon.

Electricity usage could well increase at a faster rate in Eastern Europe than elsewhere in the industrialised world with more coal fired power stations and grid interconnections with USSR, such as the one linking the West Ukraine and Hungary, helping to increase capacity.

Chapter 11

China

In 1979 China was the third largest energy consumer in the world after the USA and the USSR, mainly by virtue of a huge population of approaching one billion. By the year 2000 China will still be the world's third largest consumer and very much part of the 'super league' averaging well over one billion tonnes of oil equivalent per year, twice as much as Japan.

Such an increase in consumption, and the level of industrial growth of which it is both consequence and the the necessary adjunct is possible because of China's abundance of fossil fuels.

China's coal reserves are among the largest in the world (at 21% of the world's proved total) and it is upon use of this fuel that the country has traditionally relied, for domestic heating, industry and power generation. Accounting for 70% of energy consumption in 1979 coal as well as allowing a modest export trade with Japan, will remain the backbone of China's 'economic miracle' with production increasing to some 970 MTOE by 2000, second only to USA.

However, if coal is China's staple, oil has proved a vital and strengthening addition to the diet. Proved reserves of 2.7 billion tonnes make China the 10th largest oil province in the world - bigger than Venezuela, Nigeria or the U.K. and it is widely considered that the present high level of offshore exploration activity is likely to yield considerably more.

Although oil production has reached a plateau due to China's main oil field at Daqing having peaked, the recent decision to allow foreign oil companies to cooperate with the Chinese offshore could well lead to an increase in production within the next five years or so. Around 30 foreign surveys in the Yellow and South China seas took place in 1980 and bids for the various sectors on offer are being called for early in 1981.

Gas is the least developed of China's resources; proven gas reserves in 1979 were a modest but substantial 645 MTOE, but could be far higher.

At all events China is likely, especially in the light of the 'readjustment policy' towards conservation, to remain self sufficient in all fossil fuels and a

Chart 24 CHINA ENERGY CONSUMPTION

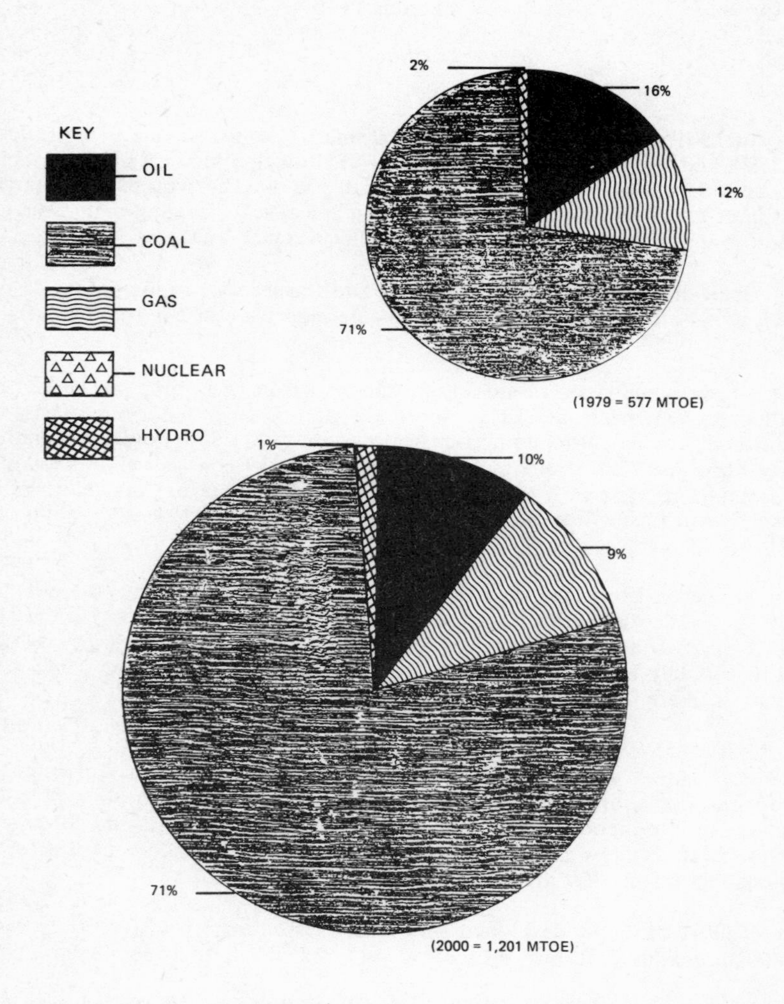

net exporter of coal for the rest of the century.

Nuclear power will probably not be seen in China during the next decade or so. Although China's scientists have been reported as recommending the construction of six nuclear power stations in order to give sorely needed extra capacity to the national grid, Peking will probably look to hydro power to improve matters, considering nuclear power stations too expensive.

Chapter 12

Canada

Like the United States, Canada is emerging from a policy of subsidising oil imports so that domestic crude prices are not exceeded. A government policy document, 'An Energy Strategy for Canada' stated that higher energy prices, i.e. raising domestic prices to international levels, would cause energy consumption to slow in relation to GDP growth but this has proved a highly contentious issue. Joe Clark, Prime Minister for only seven months, was defeated when he sought to accelerate progress towards oil self-sufficiency by, among other things, a 5.5 cents a litre excise tax on transportation fuels.

However, despite this oil self-sufficiency, to make Canada less vulnerable to supply disruptions from less stable parts of the world is still the main purpose of Canadian energy policy and certainly Canada is better placed than most to achieve this.

Coal reserves are large and coal is likely to increase its share of Canadian consumption from 10% to 16% between 1979 and 2000. It testifies to Canada's strong energy position that coal, despite its abundance, is unlikely to be used more extensively indigenously and that Canada is likely to be the world's third largest coal exporter by the year 2000 with a considerable percentage bound for Japan for use as coking coal.

Canada's gas reserves are particularly large - the 5th largest in the world and it is generally considered that the Arctic should yield a further 1-2 BTOE of gas. Gas consumption is likely to account for almost a quarter of energy demand in 2000 at around 78 MTOE but this could be considerably exceeded by production and the surplus may be either used as a feedstock for synthetic oil or exported to the US.

Canada's energy advantages do not end here: hydro electric power makes the largest contribution to energy supply of any of the world's main energy consumers at 24% in 1979 while the nuclear industry has enjoyed conspicuous success with its home grown reactor - the 'Candu'. The nuclear contribution should be raised from 4% of total energy supply in 1979 to 10% by 2000 with uranium provided from Canada's large indigenous reserves.

The Users

Alternatives to oil as an energy supplier then, are super-abundant in Canada but lowering of oil demand is not so straightforward. Canada enjoys the highest per capita rate of energy consumption in the world at 9.4 tonnes, even higher than the rate in the USA. One good reason for this is the geography of the country. Canada is larger in area than the United States and this fact, in conjunction with one of the highest levels of car ownership in the world places a heavy stress on oil's most unique usage - as a transportation fuel.

This means that Canada cannot contemplate an actual decrease in oil demand. However, conventional oil production is likely to decline in the 1980's despite the potential for offshore discovery off Newfoundland and in the Beaufort Sea.

This would all suggest the probability of increased reliance on oil imports but Canada is likely to avoid this unfortunate necessity by the rapid development and great potential of its synthetic fuels industry. Heavy oil, tar sands and other synthetic fuels should make Canada one of the world's major synthetic fuel producers by the year 2000 enabling Canada to achieve and maintain self-sufficiency in conventional crude oil into the next century.

Chapter 13

Middle East

The Middle East is the world's powerhouse but as yet it makes little use of its huge reserves. One third of the world's oil was produced in the Middle East in 1979 but it accounted for only 2½% of the world's consumption, using only 7% of its own production.

Over half the world's proven oil reserves lay in the Middle East in 1979 (56%), the same proportion as in 1970, but it is not only in the magnitude of oil resources that the advantage of the Middle East as a supplier resides.

The geological formations in the Middle East mean that the production rate per oil well is very high, and this, in conjunction with the proximity of the wells to deep water marine terminals, means that transportation and production costs are low.

The oil embargo of 1973/74 and the quadrupling of oil prices effected by OPEC was the first time the Middle East flexed its muscles and used the dependence of the Western world to its advantage. Today Western Europe receives 60% of its crude oil from the Middle East. Japan is even more dependent at 75%.

The fact that nearly all of this crude must pass through the narrow Strait of Hormuz makes supplies particularly vulnerable. Middle Eastern countries - particularly Saudi Arabia, Kuwait, Iran and Iraq, form the main proportion of OPEC accounting for 68% of OPEC production in 1979.

In addition, Iran, Saudi Arabia, Kuwait, Qatar and Iraq possess substantial gas reserves (Iran has the second largest gas reserves in the world) accounting for over 20% of world proven gas reserves.

OPEC

In September 1960 the Organisation of Petroleum Exporting Countries was formed at a meeting in Baghdad between Saudi Arabia, Venezuela, Iran, Iraq and Kuwait with the aim of unifying the petroleum policies of the member countries and determining 'the best means for safeguarding the interests of member countries, individually and collectively'.

These five countries were later joined by Qater (1961), Indonesia and Libya (1962), Abu Dhabi (1967), Nigeria (1971), Ecuador (1973) and Gabon (1973 associate member 1975 full member). There are thus 13 OPEC members of which six are Middle Eastern States, four are African, two South American and one Asian.

OPEC accounts for 67% of the world's proven oil reserves and in 1979 made up 48% of world supplies. Apart from the Middle East, 18% of OPEC production comes from Africa, 9% from South America and 5% from SE Asia.

Looking ahead there are three basic questions:

* Will world demand for OPEC oil increase?

* Can OPEC meet that demand?

* Will OPEC meet that demand?

World demand for OPEC oil will increase. By 1990 it will have risen from 1,430 million tonnes to some 1,575 million tonnes. Even with the stringent efforts now being made in the industrialised world to cut down oil use by greater efficiency, by substitution of other fuels and by energy conservation, economic growth, especially in the less developed countries makes increased demand over the next decade unavoidable. Similarly, increased oil exploration efforts worldwide will slow but not turn the tide. If there is not to be a shortfall, OPEC will need to supply the world with some 150 MTO per year more by 1990. However, this does not mean that an increase in OPEC production of only 150 MTO is required. During the next twenty years and beyond, world oil supply problems will be exacerbated by an inexorable and inevitable increase in domestic consumption within OPEC countries. With almost unlimited access to energy and vast revenues from oil exports, the OPEC countries are beginning to expand their economies. This is a process which must accelerate as size increases ability to absorb capital.

Between 1979 and 2000 oil consumption in OPEC will increase by around 570 million tonnes per year, while gas consumption will quadruple. By 2000 Saudi Arabia, Venezuela, Iran and Indonesia will be major energy consumers with per capita consumption levels on a par with some of today's industrialised countries.

The increase in OPEC domestic oil consumption means that in order to supply the world with 150 million tonnes per year more oil in 1990 OPEC production will have to increase by nearly 300 million tonnes per year.

By 2000 alternative energy sources will be lowering world demand for OPEC oil below the 1979 level, but the 'joker in the pack' - OPEC domestic consumption increase - will mean that an OPEC production level of over 2 billion tonnes a year will be required if demand is to be supplied.

Can this level be achieved? One of the OPEC Secretariat's economic analysts,* going against a number of statements from other sources that OPEC production had already peaked, stated that although Algeria, Ecuador, Gabon, Indonesia, Nigeria and Venezuela would not be able to raise present production levels, other OPEC countries have sufficiently high ratios of reserves to production to meet the world's growing needs.

OPEC production in 1979 was 1,544 million tonnes. Looking ahead to when Iran and Iraq can produce again (present OPEC production capacity is about 1,750 million tonnes). The OPEC economic analyst saw a level of about 2,200 million tonnes as possible by around 1990.

Such a level of production, will, in the opinion of this study not be necessary. However, in view of the fact that proven OPEC reserves alone could sustain 1979 product until about the year 2020 while estimated recoverable oil in the Middle East alone would last for over 50 years, the possibility of a level of production of up to 2,000 million tonnes per year from OPEC must be considered feasible.

Whether OPEC will increase production, even above today's level has been the subject of much debate. The most frequently proposed argument against an increase in production has been that, in view of the tremendous revenue surpluses being amassed within OPEC and the impossibility of absorbing these sums into economies already paralysed with unspent money, many OPEC countries would be quite happy to reduce output.

Absorption means the transformation of wealth acquired into some permanent element within a country's socio-economic infrastructure. If this cannot be achieved then the country concerned is, in essence, frittering away a finite and diminishing resource. The runaway accrual of liquid capital is not sufficient incentive to deplete one's oil resources.

Could the increased revenues from increased production over the next two decades be absorbed? Present and projected levels of economic growth within OPEC countries (of which increased energy use is a symptom) strongly suggest that increased revenues will be necessary.

* Oxford Energy Seminar 1980 - M.V. Samii

The Users

This study shares the view that the OPEC high absorbers will absorb increased funds much sooner than expected. This view is supported when one considers that in 1974 OPEC was only able to absorb 43% of revenues while by 1978 this proportion had risen to 90% while even after the price increases of 1979 the $65 billion OPEC surplus represented only 25% of total revenue.

Chapter 14

Australia

A country which is large enough to be a continent had a good chance of being well endowed with primary fuel resources and this is the case with Australia. There is enough coal and gas to make Australia a significant exporter of both while uranium deposits, which will not be needed for domestic use within the next decade or two, are the fourth largest in the world. The trouble is that being so large means a particular need for transport fuel. Transport fuel means oil and Australia's oil reserves are relatively low.

In 1979 oil accounted for 45% of Australia's energy needs and although the growth in usage of alternative fuels will lower this proportion considerably, actual demand will increase to 1990 and by 2000 will still not have fallen below the 1979 level. Only a great success in present offshore exploration efforts will avert a growing need for imported crude oil. Although some of this demand will eventually be offset to an extent by coal liquefaction and by exploitation of Australia's shale oil deposits there will be very little contribution by this means up to 1990.

Conservation of oil and its more efficient use is particularly important in Australia if the proportion of energy use accounted for by oil is to be lowered. The main methods of achieving this are the replacement of oil fired power stations by coal and increasing the proportion of small cars on the roads - (during the past 15 years the proportion of cars with less than two litre capacity has increased from below 30% to above 50%).

Oil production decline in Australia will be compensated for by dramatic growth in both coal and gas production over the next 20 years.

Australia has the fourth largest recoverable reserves of hard coal in the world. During the next 20 years coal will become Australia's main fuel accounting for 40%* of energy consumption in 2000 but most significant and impressive growth will be in coal exports. By the year 2000 Australia should be the second largest exporter of coal in the world, exporting over 100 MTOE.

* Australasia

The Users

This level of export success depends on a number of factors. Firstly, the Australian coal industry will need a huge inflow of capital if resources are to be developed quickly enough. However, this should be achieved with the help of foreign investment - for example there has already been substantial investment in Australian coal from the major oil companies.

Secondly, the necessary infrastructure must be established. Australian coal production costs are highly competitive due to the abundance of deposits close to the surface (particularly of black steam coal for power stations which is the main growth area in the coal trade).

Transportation costs are at present high but these may be easily reduced by the use of larger purpose built coal tankers. Of the 26 MTOE of coal exported in 1978-9 66% went to Japan and only 18% to the EEC. The next 20 years are likely to see significant efforts to increase import penetration into Western Europe.

Japan will also be the recipient of almost all Australia's LNG exports to 1990 (8 MTOE). Natural gas will be the chief energy growth area in Australia over the next 20 years as far as domestic consumption is concerned accounting for some 25%* in 2000.

One of the most significant decisions for Australia's energy economy was taken late in 1980 when the Australian government decided to accelerate the development of uranium mining. Australia has about 16% of the Free World's easily available (up to $80/kilogram U) uranium and although the slowdown in nuclear power development has meant that there is at present a glut of uranium on the world market, Australia will be ready to become one of the world's leading suppliers as consumption increases and demand and prices follow suit.

The announcement by the Government of the conditions under which Australian uranium may be reprocessed to obtain plutonium is a further confirmation of Australia's determination to become a major uranium exporter.

* Australasia

Chapter 15

Latin America

Apart from Mexico and Venezuela, both of which have large oil and gas reserves, South America is relatively poor in proven fossil fuel resources.

Proved coal reserves are, at present, about as large as those of the U.K. while gas reserves are only slightly larger than those of Western Europe. South America's oil reserves made up 9% of the world's proved total in 1979 but the figure has doubled over the past 10 years - mainly as a result of discoveries in Mexico.

Over the period 1979-2000 Latin America will increase its energy demand by over 100% and its share of world energy consumption from 4.6% to 5.5%. This significant increase however is not a general one. As in the other less developed regions - Africa, South Asia and South East Asia - it is the countries with indigenous fuel which expand while those without are likely to suffer actual contraction in per capita consumption.

Thus more than half of the growth in energy demand in Latin America will occur in two countries - Venezuela and Mexico. With oil reserves of about 2.5 billion tonnes, Venezuela has the eleventh largest proved conventional reserves in the world and was, in 1979 the largest oil producer in Latin America and sixth largest in the world at 125 million tonnes.

Venezuela has however suffered considerable decline in production over recent years, in 1970 production came to 195 million tonnes, 70 million tonnes higher than in 1979.

For forty years, between 1929 and 1969 Venezuela was the world's largest exporter of crude oil. The decline in production from the peak of 1970 was not because of depletion though - proved reserves have risen since 1970 - but because of the rapid decapitalisation of the Venezuelan oil industry as the oil companies pulled out their money in anticipation of the 1976 nationalisation. Venezuela is now engaged in rebuilding the oil industry and raising production. Some success has already been achieved: 1979 production was 9% up on 1978, the first increase for nine years.

A considerable exploration effort has been mounted to find new oil fields and although little more conventional light oil has been discovered, gas reserves have been doubled and a great deal of non conventional oil discovered.

The huge deposits of non conventional heavy oil in the Orinoco belt, though expensive to exploit being heavier than water and containing large proportions of sulphur and metals, are expected to yield some 50 million tonnes of oil per year by 2000.

Venezuela's exports of oil will tend to reduce even if present levels of production are raised due to the tremendous growth in domestic consumption. By 2000 Venezuela's oil demand could well be approaching 100 million tonnes if present rates of increase are maintained.

Venezuela was Latin America's main gas producer in 1979 at 15.3 MTOE which was 33% of Latin American production.

Mexico, with a population of 68 million, five times that of Venezuela, is likely to increase its energy consumption even more quickly over the next 20 years unless brakes are applied.

Mexico has, over the past 10 years, become the hydrocarbon capital of Latin America and now has the 5th largest oil reserves in the world. In 1970 Mexico's oil reserves represented 12% of Latin America's total, in 1979 - 55%. Mexico's new found hydrocarbon wealth does not end with oil; gas reserves now account for 43% of the Latin America total.

Mexico will soon overtake Venezuela as Latin America's main oil producer - although 1979 production at 80 million tonnes was 40 million tonnes lower than that of Venezuela, the 1980 figure is likely to be at 115 million tonnes or more.

A long range plan was recently announced with the aim of rationalising domestic oil consumption and not allowing oil exports to rise above 75 million tonnes per year after 1982. Part of the plan is to reduce domestic dependence on oil and natural gas which in 1980 was running at about 85% but while domestic petrol prices are kept at artificially low levels this will be impossible. The construction of a second nuclear power station is also proposed for the 1980's as part of the strategy to promote alternative fuels..

The other countries with the indigenous resources to achieve some measure of growth in energy usage over the next 20 years are Brazil, Argentina and Colombia. Argentina has the third largest oil and gas reserves in Latin America while Colombia has some oil and gas and significant coal resources which foreign investment is now helping to develop.

Brazil, the biggest country in Latin America in both population (117 million) and area was the largest oil consumer and importer in 1979 accounting for almost a quarter of Latin America's oil consumption but only 3% of production. But Brazil's prospects cannot be assessed using the criteria which applies to the less developed country; by virtue of the level of industrial development and the vigorous if mercurial nature of its economy Brazil will be able to maintain growth in energy usage without seriously damaging economic growth.

The cost of oil imports - in 1979 just under half Brazil's export earnings, is the main problem for Brazil. The aim is to reduce this by increasing domestic production of oil, by the substitution of synthetic fuels made from sugar cane alcohol and coal and by the development of hydro electric and nuclear power to replace oil in industry.

Oil production could be raised by some 10 million tonnes per year but it is considered unlikely that domestic production will ever exceed 20 million tonnes per year. Transport in Brazil is largely dependent on petroleum fuels because its rail system does not cover many areas. The changeover to alcohol as an automotive fuel is moving quickly in Brazil. In 1980 about one quarter of all cars produced will be designed to run on alcohol while most cars are already using petrol mixed with 20% alcohol.

Though not well endowed with oil and natural gas Brazil has the best hydro electric power potential of any country in the world - estimated at 213,000 MW, around four times the capacity of U.K.'s national grid. At present hydro electric power is responsible for around 90% of electricity generated (oil 5%). Electricity demand, the main growth area in energy consumption in Brazil, will account for an increasing proportion of total consumption helped by the huge hydro electric projects now underway. The Itaipu project on the Parana river, largest project of its kind in the world, will, when completed, produce 12,600 MW - as much as 10 large nuclear power stations while the flow of water through the 18 turbines will be 150 times that of the River Thames.

By the year 2000 22% of Latin America's energy will be produced by hydro electric power.

For the other Latin American countries, oil dependence coupled with lack of developed indigenous resources will prove an insuperable obstacle to real economic growth over the next 20 years without large scale foreign investment in alternative fuels and technologies.

Chapter 16

South Asia

The countries comprising this regional grouping are Afghanistan, Bangladesh, Burma, India, Pakistan and Sri Lanka.

Economic growth between 1970 and 2000 is likely to average around 3.8% - approximately the same as energy growth. Unless population growth can be controlled this will mean little or no improvement in per capita consumption - the worst in the world at 0.1 tonne of oil equivalent per inhabitant (87 times lower than per capita consumption of energy in North America).

Apart from India's large coal reserves and modest oil reserves and Pakistan's modest gas reserves, proved fuel reserves are low. India is the main energy consumer and the main fuel consumed in South Asia is therefore coal (57%). However, oil dependence is very high outside India and the inevitable growth in oil consumption from 37 million tonnes in 1979 to 59 million tonnes or so in 2000 will prove very expensive to these poor countries.

In India coal will be increasingly used in the home, in industry and for electricity generation while 10,000 MW of nuclear power are planned by the end of the century.

Chapter 17

South East Asia

South East Asia is, with a few exceptions, one of the most buoyant parts of the 'Third World'. In the next 20 years SE Asia's share of world energy consumption will increase from 2.6% to 3.2%. Growth within the area will be mainly concentrated in the ASEAN (Association of South East Asian) nations - Indonesia, Malaysia, Singapore, the Philippines and Thailand. In contrast to South Asia, South East Asia, which is more highly industrialised is heavily dependent on oil which supplied 64% of energy needs in 1979. Oil and non-oil energy reserves are, however, good. Indonesia has the 13th largest oil reserves in the world at 1.3 billion tonnes and is a major exporter, in 1979 using only 25% of production indigenously.

Malaysia also has fairly substantial oil reserves and in 1979 exported over 15 million tonnes. Both Indonesia and Malaysia have substantial and growing gas reserves but, at present, gas consumption in the area is fairly low, though this will increase from 4% to 10% of total energy consumed by 2000.

Coal use will also increase significantly over the next two decades, and there will be a significant expansion of nuclear power though this will be largely concentrated in the countries least well endowed with fossil fuels - South Korea and Taiwan. Development of these alternative fuels will mean that oil usage will account for 50% in the year 2000, as opposed to 64% in 1979.

Indonesia, Singapore and the Philippines are the main oil consumers and in all three (despite Indonesia's reserves) efforts are being made to reduce oil dependence. Indonesia has considerable low grade coal reserves which could be used to generate electricity, possibly for export if an ambitious scheme to link its own electricity grid with that of Malaysia, Singapore and Thailand goes ahead. The Philippines is beginning to produce oil and could be supplying 20% of its own demand by 1981.

The Philippines, like Indonesia has large undeveloped coal resources and the added bonus of abundant geothermal resources which could reduce the oil contribution to electricity generation from its present level of 70% to 40% by 1985.

The Users

Thailand is even more heavily dependent on imported oil which accounted for over 80% of energy consumption in 1979 but now plans to develop domestic natural gas and coal resources in order to reduce oil imports by 40% by 1986.

The South East Asia energy exporters' main market is Japan. As Indonesia's domestic needs expand and her oil exports fall, natural gas exports from Indonesia and other SE Asian countries to Japan will increase.

Chapter 18

Africa

Africa had the second lowest* per capita consumption of energy in the world in 1979 at 0.3 tonnes oil equivalent per capita. Comparatively speaking the degree of oil dependence was fairly low in 1979 (at 44% of total energy consumption) but the picture is distorted by the high coal consumption in South Africa and for most countries oil dependence is high.

During the years 1979-2000 energy consumption in Africa will increase fourfold but the increase will not be a uniform one as a large proportion will be accounted for by the OPEC countries in Africa - Nigeria, Algeria, Libya and Gabon - which between them will make up some 37% of Africa's energy consumption in 2000 and 65% of oil consumption.

In 1979, 86% of Africa's proved oil reserves were in the three OPEC countries Libya (41%), Nigeria (31%) and Algeria (14%). Both Libya and Algeria's oil reserves are, however, lower than they were 10 years ago and reserves within the continent of Africa as a whole are 23% lower than in 1970.

These three countries also have the largest proved gas reserves in Africa. Algeria has the 4th largest gas reserves of any country in the world (4% of total proven gas reserves) and is the world's leading LNG exporter. The discrepancy between the production and the consumption of both oil and natural gas in Libya, Nigeria and Algeria is of an extremely high order - only 4%, 4% and 8% respectively of oil production being consumed indigenously but this ratio will change drastically over the next twenty years with no significant increase in production with the result that there will be far less oil for export to the world's industrialised countries.

By far the main energy consumer in Africa is South Africa which accounts for 24% of oil consumption. South Africa, one of the world's major coal provinces, is also a significant consumer of coal. The country's political isolation and lack of indigenous oil has made coal liquefaction a major technology in South Africa.

* of the regional divisions used in this study

The Users

Fuel production in countries other than South Africa and the OPEC countries is extremely low. Egypt, with the second largest population in Africa (40 million as against 74 million in Nigeria) is the second largest oil consumer, accounting for 18% of Africa's consumption and is a net exporter to the tune of some 14 MTOE per year, however reserves are modest and declining and unless the present major offshore search proves successful Egypt is unlikely to remain self-sufficient in oil to the end of the century.

Gabon and Angola are also net oil exporters but proved reserves are small.

The rest of Africa suffers from the necessity to import high cost fuel, lacking the resources to develop indigenous resources. Often although reserves exist which could meet indigenous needs, these are not large enough to attract foreign investment in their development. In addition, unstable political regimes are a considerable disincentive to risking capital.

Oil has been discovered in Benin, Chad, Niger, Ivory Coast and Sudan but has not, as yet, been exploited and these countries continue to import oil.

Considerable undeveloped coal reserves also exist in Africa. Botswana and Swaziland are likely to be producing coal in significant quantities by the end of this decade, while Egypt, Niger and Tanzania have reserves but, so far, are not producers.

Over the next decade the fuel which will account for the largest proportion of growth in energy consumption in Africa, will be natural gas - which will account for 11% of total consumption in 1990 as opposed to 6% in 1979, a fourfold increase. However, most of this increase will be in Nigeria, Libya and Algeria, which in 1990 will account for 80% of consumption.

Chapter 19

Less Developed Countries

It was predicted at the World Energy Conference that the world's population would increase by about 50% by the year 2000. This is at a slower rate than the rate at which energy consumption will increase therefore average per capita consumption of energy will be about 0.3 tonnes of oil equivalent higher in the year 2000 than it was in 1979.

About 90% of the world's population growth to the year 2000 will be within the developing countries, but this will be matched by a 178% rise in energy consumption in the LDC's such that there will be an actual rise in per capita consumption from 0.4 to 0.6 tonnes of oil equivalent per year.

This increase, though appreciable in percentage terms, when viewed in a world context, is of miserable proportions. The world average per capita consumption is likely to be some 1.3 TOE higher in 2000 than per capita consumption in the LDC's and will also experience a higher actual growth (0.3 TOE as against 0.2 TOE) Per capita primary energy consumption in the LDC's should also be seen in relation to that of the industrialised countries. In 1979 the average inhabitant of North America used 8.7 TOE,eighty seven times more than the average inhabitant of South Asia. Taking individual countries as examples the discrepancy is even more marked - according to the Brandt Commission, one American uses as much commercial energy as two Germans or Australians, three Swiss or Japanese, nine Mexicans or Cubans, 16 Chinese, 19 Malaysians or Indonesians, 109 Sri Lankans, 438 Malians or 1,072 Nepalese. Energy Commission paper 13 stated that the average Canadian's energy consumption in 1976 (9.1 TOE) would have been sufficient to support 130 Nigerians (0.07 TOE per year).

Even the increase of 0.2 TOE in per capita consumption 1979-2000 in the LDC's does not mean any increase in standard of living in most of the countries considered. Almost all the growth in consumption will be in OPEC countries and other energy resource rich countries such as Mexico.

Of the 121 countries which the World Bank designate 'developing world' 17 countries account for 85% of all commercial energy consumed.

The Users

In 1979 the LDC's of the Middle East, Latin America, SE Asia, Africa and South Asia accounted for 13% of total energy consumption and 16.2% of oil consumption. By the year 2000 the level of consumption will have risen to account for 20% of total energy use and 30% of total oil consumption.

The growth in energy use within the oil exporting countries is the most significant factor in this growth. Thus the main growth areas are the Middle East and Africa where most of OPEC's resources are concentrated. These two areas will increase their share of LDC energy consumption from 28% in 1979 (12% Middle East, 16% Africa) to 45% in the year 2000 (20% Middle East 25% Africa). The main consumption area will, however, remain Latin America which will account for 27% of total LDC energy consumption in the year 2000.

Latin America and SE Asia will achieve reductions in oil use by the development of alternative energy sources, for example, hydro electric power in South America, but both will remain heavily dependent on oil. South Asia, the region with the highest population and lowest per capita consumption, though less dependent on oil than the other regions,will suffer more from oil import prices by virtue of its poverty.

World oil price rises have contributed to the plight of many LDC's whereby a higher and higher percentage of export revenues must be used to buy oil products.

Chancellor Schmidt at the World Energy Conference used the example of Costa Rica:- in 1972 28 kilogrammes of bananas bought one barrel of oil. In 1980 a barrel of oil was bought for the revenue on export sales of 420 kilogrammes of bananas. Other examples given were that one third of India's export revenue went towards oil purchase while in Brazil the percentage was 40% and in Turkey 60%.

Oil prices have increased ten fold in the last decade and will certainly double again by 1990. In 1980 the LDC's are likely to have to pay $35 billion more for their oil than in 1979. Meeting this level of expenditure is a tremendous burden for countries with low borrowing capabilities and high debts and unless significant funds are made available many will not be able to pay.

At the Munich World Energy Conference Chancellor Schmidt called for OPEC members to give more aid to developing countries than the $5 billion allocated in 1979, out of a probable trade surplus of some $120 billion in 1980. In 1979 funds from the Organisation of Economic Cooperation and

Development Countries paid for one third of the developing world's oil. OPEC's aid to the developing world has fallen, as a proportion of GNP, from a peak of 2.71% in 1975. The OECD countries have contributed some 0.34% of GNP over the past decade.

The reason why the LDC's energy consumption must grow at levels of up to 7% per annum and that even in the most depressed regions (e.g in South Asia) it will grow at rates far higher than that in the industrialised world, is that the early phase of industrialisation is highly intensive e.g. the development of heavy industry and transport systems.

In addition a switch is taking place away from traditional non commercial fuels such as wood, which are anyway running out fast in many countries.

The point is that energy consumption must increase if drastic declines in living standards are not to occur but the only way this can practically be achieved for the 90 (approximately) LDC's which must import oil,is the speedy development of indigenous resources. The World Bank considers that aggressive exploration in these countries plus use of techniques to recover heavy oil could raise production from the oil importing developing countries to a level of some 240 million tonnes per year.

In order to achieve this, however, and increase the falling level of exploration drilling, private oil companies, as opposed to state concerns, will need to be encouraged to invest. This is already happening in Argentina, Brazil, Pakistan and Turkey.

However, this alone would not be enough to develop the reserves in the oil importing developing countries, which have been estimated at an ultimately recoverable 7 billion tonnes. This is because, in many cases, the promise of return on risk capital is not good enough. To overcome this problem the World Bank has called for an 'expanded energy programme in the oil importing developing countries' (OICDs) involving the investment of $450 billion-$500 billion over the next decade - an amount which is lower in yearly terms than the OICD s oil import bill in 1980.

The development of indigenous resources would not be restricted to oil. Natural gas is seen as a good energy bet contributing a possible 100 MTOE per year to OICDs by 1990. Much of this would be associated natural gas - the gas occurring with petroleum deposits, which was, in the past, generally flared off at the well head. Today, on average, about 60% of associated gas is used. The problem is that the pipelines and distributing systems necessary for domestic use of gas are very expensive - a gas pipeline is between five and eight times as expensive as a pipeline carrying oil of the same energy content.

Projects to enable gas to be used domestically are at present underway in Bangladesh, Egypt, Thailand and Tunisia.

Coal is, at present, mined in 29 of the developing countries and production could increase to around 240 MTOE by 1990 with the main growth in Brazil, Colombia, India, South Korea, Mexico, Romania, Turkey, Vietnam and Yugoslavia. The World Bank has called for investment totalling $175 billion to $350 billion over the next decade for mines, transportation facilities ports and ships in order to meet growing demand.

When considering fuel use in the developing world it should be remembered that only primary fuel consumption is statistically recorded and that about 25% of all energy consumption in developing countries is provided by fuels such as wood, charcoal, crop residues and animal dung.

More than 1,500 million people in developing countries depend on wood for cooking and keeping warm. In Africa the contribution of trees to total energy use is as high as 58% while in SE Asia and Latin America it is 42% and 20% respectively[1]. The effect of such intense demand has caused extensive deforestation. Around one fishing centre in the Sahel region of Africa, where the drying of 40,000 tonnes of fish consumes 130,000 tonnes of wood every year deforestation extends as far away as 100 kilometres[2] while fuel wood is now so scarce in the Gambia that gathering it takes 360 woman days per family[3].

Lack of wood makes for increased use of animal dung and crop residue for fuel and a valuable fertiliser is thereby lost.

The World Bank considered in its report that it would be necessary to plant 50 million hectares of trees - a fivefold increase over the current rate - in order to provide the domestic cooking and heating fuel which will still be necessary in many developing countries.

1 FAO 1978 The State of Food & Agriculture 1977, FAO Rome
2 FAO 1971 Environmental Aspects of Natural Resource Management - Forestry
3 FAO 1978 Forestry for Local Community Development, FAO Forestry Paper 7

Use of wood and animal wastes as fuels is the traditional application of what is now referred as 'biomass' - a renewable energy source. A newer application which will prove of use in some developing countries is the manufacture of alcohol by the fermentation of sugar, for use as a transport fuel. However, this will not be practiced generally because in essence, it is using food as fuel and many LDC's have food as well as energy problems.

Geothermal power will also make a limited contribution in LDC's which enjoy the tight geological conditions - El Salvador, Indonesia, Kenya, Mexico, the Philippines, Nicaragua and Turkey are all potential users of the heat of the earth's crust.

Geothermal heat may be used either for district heating or for electricity generation.

The installed electricity generation capacity in the developing countries is 241 gigawatts (about five times that of the U.K. alone and 12% of global capacity).

The World Bank report estimates that this will increase to 524 gigawatts by 1990 with the use of oil as a power station fuel falling from 37% to 25%, geothermal contributing 2.3 gigawatts and nuclear power increasing its contribution from 1.4% to 7.3%.

In 1979 developing countries with nuclear power stations included Argentina, Brazil, India, South Korea, Taiwan and Pakistan while plants are under construction in Mexico, the Philippines, Rumania and Yugoslavia. The World Bank states that by 1990 Egypt, Portugal, Thailand and Turkey may also have nuclear power stations.

The World Bank lending programme, as well as the increased contributions that OPEC could make to the reduction of the OIDC's debilitating dependence on expensive, imported oil through utilising indigenous resources, is vitally important. It is important not only on humanitarian grounds - otherwise standards of living will fall even lower - but to help maintain world stability by reducing the chances of a worldwide dutch auction with oil at grossly inflated prices at the centre of the scrum and the LDC's at the back of the queue.

Part Four
ASPECTS OF THE ENERGY FUTURE

Chapter 1

Synthetic Fuels

Synthetic fuels, or 'synfuels' may be defined as either liquids or gases which may be substituted for oil products or natural gas. Since they may be obtained from coal, heavy oil, tar sands, oil shale or from vegetable matter the problem with synfuels is not one of reserves but of the cost of their production or manufacture.

Recent oil price rises mean that a number of synfuels have crossed the cost effectiveness threshold. The American Petroleum Institute estimated in a report in mid-1980 that coal based gas could be manufactured and sold for the equivalent of between $30 and $36 a barrel of oil, while the cost of shale oil and liquids from coal was put at $30-$50 a barrel.

In view of the fact that Saudi Arab Light Crude rose from $1.75 to $32 a barrel during the decade 1970-1980 and that this rise took place despite the fact that world crude oil supply was in surplus for all but two short periods during that time means that the decade 1980-1990 will price many synfuels and also many other alternative energy sources, firmly into the market.

The United States and Canada will take the lead in synthetic fuel production and be responsible for some 78% of consumption in 1990. Brazil and Venezuela also have significant potental for synfuel development.

The United States will lead the world in synthetic fuel consumption, firstly because it has substantial reserves of synfuel source materials but secondly because the financial resources exist to spend the huge sums necessary to establish synfuel facilities and sustain investment over long lead times. In addition the US has the incentive: to reduce its expensive and weakening dependence on oil imports.

This combination of reserves, money and incentive in the United States has enabled the passage of the Energy Security Act (June 1980) under which a Synthetic Fuels Corporation would be set up with the authority to spend $88 billion by 1992 to create a new synthetics industry.

The United States has enough coal and oil shale for the manufacture of

Chart 25

LEAD TIMES IN ENERGY SUPPLY *

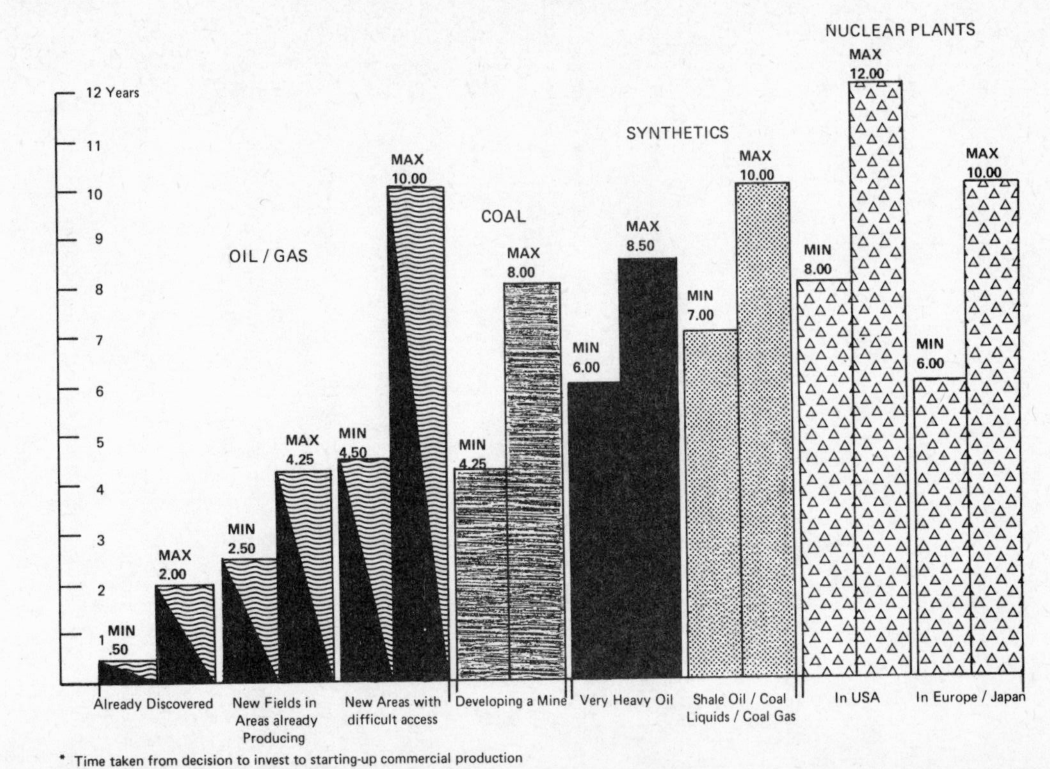

* Time taken from decision to invest to starting-up commercial production

750 million tonnes of synfuels per year for a period of about 180 years (of which about 55% would be from oil shale and 45% from coal). The amount of investment required to build the facilities to produce at this rate would cost more than $3 trillion dollars (trillion = million million) an amount which, despite its being spread over a number of years, would represent a titanic commitment of resources.

In addition to the cost, the environmental effects of such large scale developments and also the huge amount of water needed are the main problems. The latter problem would be a major concern in the US since most of the deposits are in the West - Wyoming, Colerado, Utah. There are shale oil reserves of varying quality in Brazil, Estonia, China, Scotland, England and Germany and elsewhere.

Tar sands, like oil shales are very expensive to recover and process. As with the modern coal industry, surface mining is the most economical method of recovery. A tar sands mine might be three miles long and half a mile wide and would move ponderously cross country at about 700 feet per year producing.

The main deposits of oil sands where oil production is underway are in the Athabasca area of Canada and the Orinoco basin region of Venezuela.

Oil from coal (coal liquefaction) is the other main synfuel. Like the conversion of oil shale which was practised in Scotland in the late 19th century coal liquefaction is not new - Germany produced gasoline from coal during World War II by the Fischer-Tropsch process, a technique used today in South Africa to produce around 10 million tonnes of oil per year.

One synfuel is also a 'renewable' energy source - this is alcohol produced by biomass. Alcohol is already used extensively in Brazil to replace gasoline either totally or partially as a transport fuel (see 'South America' section).

Sugars, starches and celluloses may be fermented to produce ethyl alcohol or ethanol. Brazil, with two sugar crops per year and agricultural land to spare is the best endowed country in the world for the production of alcohol but the use of biomass - surplus agricultural products such as bananas or corn - could reduce oil consumption for transport. Alcohol may either be used as a sole fuel or mixed with petrol (gasohol). In the US a 10% concentration of alcohol in petrol is offered and a programme begun under President Carter will produce 500 million gallons of alcohol for blending with other fuels by the end of 1981. Persuading people to use 'gasohol' does not appear to present problems. It has a higher octane level than petrol and can increase fuel efficiency.

The Future

The use of biomass fuel to reduce oil dependence would be particularly applicable in agriculturally strong developing countries. The World Bank mentions Mali, Sudan and Thailand for their large biomass production potential.

Use of biomass for energy could be controversial where countries were self-sufficient in neither food nor energy. However the World Bank points out that world energy prices are expected to rise more quickly than world food prices over the next 10 years. This could create the odd situation that it was more economical for a country to reduce its fuel imports and increase its food imports by using more and more of its agricultural production to produce energy and less to feed itself.

Peat should also be mentioned in the context of synthetic fuel feedstocks. According to a paper presented to the '7th Energy Conference and Exposition' (Washington DC, March 1980), 250 billion tonnes oil equivalent of peat reserves exist in the world - more than half of which is in the USSR. Although peat has a lower calorific value than coal it has been suggested in some quarters that peat could be preferable to coal as a synthetic fuel feedstock. Peat also has the advantage of being closer to the surface than coal, therefore easier to produce.

Chapter 2

Oil Prices

As the Brandt Commission pointed out in its chapter on energy, the economics of oil have been controversial since the first commerical oil strike in Pennsylvania in 1857.

Over the past ten years oil prices have risen from $1.35 per barrel to $32.00 per barrel (Arab Light - 34). The increase in price of some other oils has been greater than that for Arab Light - 34, the price of Libya's Es Sider* - 37 has risen from $2.01 to $36.78 while Nigeria Bonny* - 37 stood at 37.02 in mid-1980 - an increase of $34.92 over the 1970 level. In real terms too the increase has been a considerable one; in 1970 dollar terms oil prices per barrel in January 1980 would have risen to over 18 dollars - a tenfold increase. However, the cataclysmic impact that oil price increases have had, both on the world economy and on energy prices generally, have not been primarily due to the amount prices have risen but to the suddeness with which they have risen.

In essence oil price increases during the seventies were necessary and overdue. During the 1960's oil was cheap and consumption in both industrialised and some developing countries was able to increase without the restriction of a price reflecting future scarcity. In fact oil was 25% cheaper in 1975 in relation to the price of other goods than in 1955.

Viewed rationally oil prices in 1980 should certainly be no lower, if they were a number of alternative fuels would not be economic and most countries now accept that realistic oil prices which reflect the value of the product on the world market are one of the most effective mechanisms for conserving and using it efficiently. The chief cause of economic hardship has been the way in which oil prices have increased. The problem has been the sudden leaps in price and the huge transfer of wealth from importer to exporter that this has entailed. The changes in the current account balance of the OPEC nations when compared with that of the industrialised countries look like a roller coaster ride when plotted on a graph but the sums concerned are billions of dollars. In 1972, before the oil crisis, the current account of the industrialised countries stood at around $10 billion while OPEC's was nearer four billion dollars. In 1974 the position changed with the OPEC surplus topping $60 billion and the industrialised countries acquiring a deficit of over $30 billion.

Chart 26

CRUDE OIL PRICES 1970 - 1980

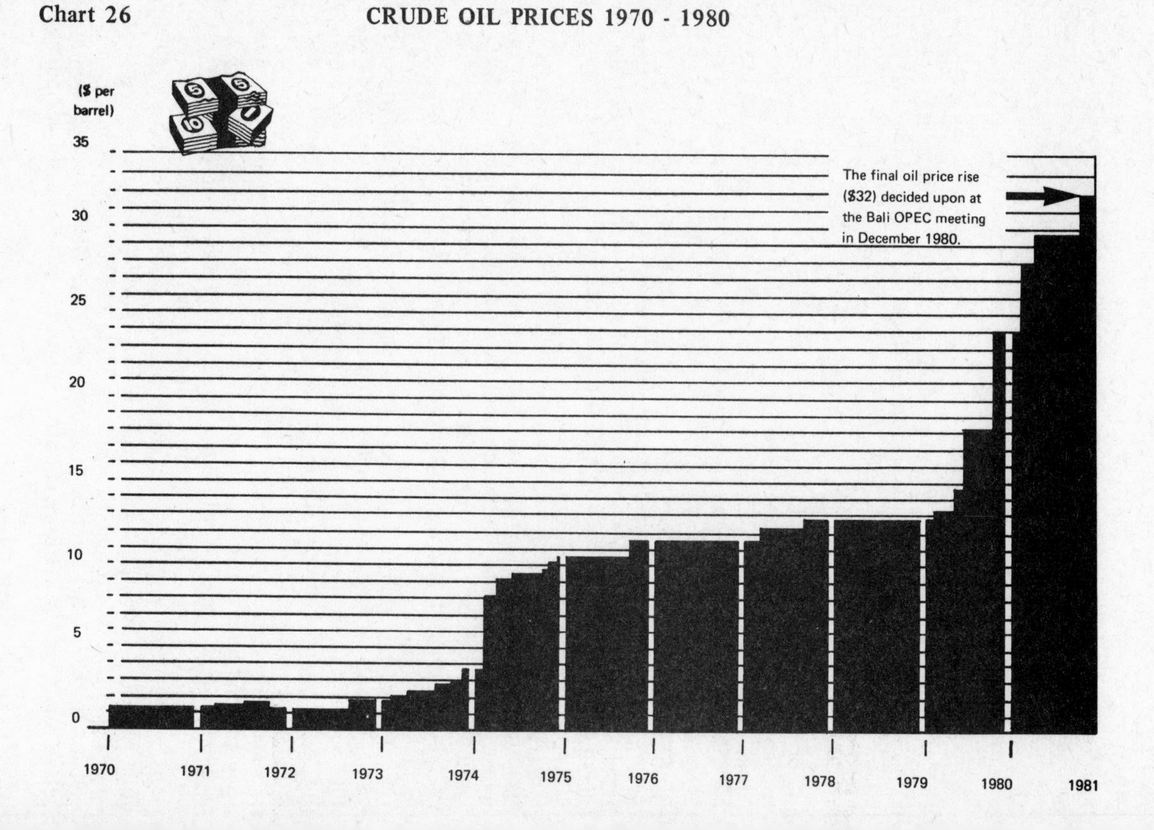

By 1978 the OPEC surplus had fallen to around $12 billion and the industrialised countries had retrieved their economic situation to enjoy a surplus slightly greater than OPEC's. The 1979 price rise caused the second traumatic upheaval. The OPEC surplus in 1980 will touch a staggering $120 billion with the industrialised countries about $84 billion in current account deficit.

According to David Howell, U.K. Secretary of State for Energy, speaking at a meeting in July 1980, 'a 10% oil price rise results in a $25 billion deterioration in the OECD current account position in a full year' while also reducing OECD GNP by 0.4% after one year and also affects inflation by adding an estimated 1% to OECD prices after one year.

OPEC's strategy for the avoidance of these shocks to the world's economy is a system of indexing.

Dr. Rene Ortiz, OPEC's Secretary General, said at the World Energy Conference in Munich that an indexed price structure, adjusted for inflation and currency fluctuations would protect oil tariffs in real terms and bring stability and predictability to world energy markets. But Dr. Ulf Lantzke, Executive Director of the International Energy Agency, considered that an indexing system might not preclude major oil price rises and that a real danger would exist that consumers would suffer indexed price rises during times of slack demand and in addition, sharp increases caused by a tight market during times of shortage.

It is interesting to note that over the last ten years there have only been two occasions, of 18 months' duration in all, when there has actually been an oil shortage, yet crude prices have still risen although it is true that most of the increases came in two big jumps - the first in 1973/4 when OPEC first increased the posted price of oil by 7% (October 1973) then (in January 1974) more than doubled the price of crude; the second in 1979 when OPEC price unity broke down and prices rose, between March and December by over $10 per barrel.

Oil prices will rise between 1980-2000 in real terms, probably by not less than 2% per annum on average and possibly by no more than 3% per annum. However, the pace of increase is impossible to forecast.

The latest increases in oil prices - agreed at Bali in December 1980 threaten to raise the average price of crude oil to $35 a barrel.

At the meeting a new market ceiling for oil of the Arab Light type was set at $36 a barrel while for higher quality crude e.g. from Libya the ceiling

Chart 27

WORLD CURRENT ACCOUNT BALANCE

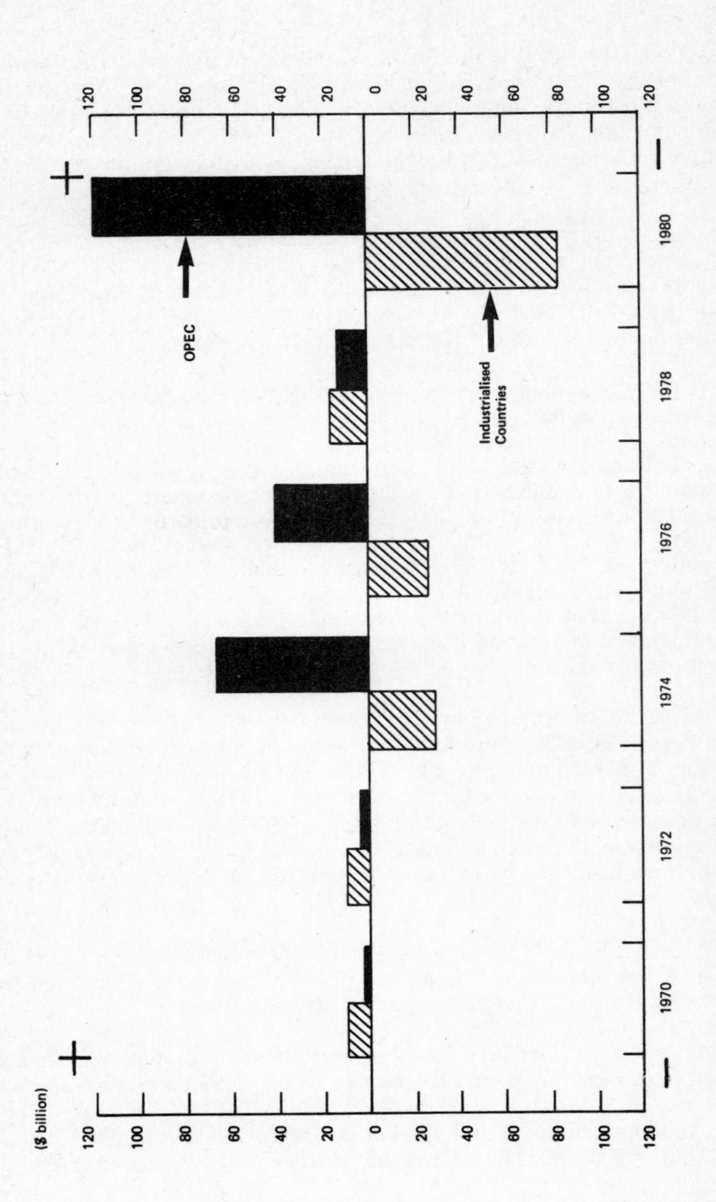

is $41.

Libya has tended, since the last breakdown of OPEC price unity in late 1979, to be one of the 'pricing hawks'; Saudi Arabia formerly the holder of the crude oil agreed by OPEC as the reference price against which the tariffs of all the other crudes were set ('marker crude') has always sought to exercise a moderating influence on price increases and to try to re-establish unification of price.

At the Bali meeting Saudi Arabia agreed to raise the base price of its Light Reference crude to $32 per barrel.

The next OPEC meeting in the Summer 1981 in Switzerland could bring a return to unified pricing. This is its turn would make regular increases - indexed to the Western World's level of economic growth, possible. Such a system would certainly benefit the West and smooth the path to world economic growth. But with a market as volatile as the oil market,dependent as it is on a stable Middle East, the part of the globe where stability is least likely, the odds against uniform, gradual and regular price movements are very long indeed.

Chapter 3

Conservation

The amount by which world energy conservation measures could reduce world energy consumption in the year 2000 has been estimated at between 20% and 30%*. This would be if all possible measures were implemented of which a proportion would require considerable capital investment.

10-15% savings of energy are relatively easy to implement by such measures as better insulation, improved heating controls and some change in work practices. However, to progress beyond this, long lead times and a great deal of money are necessary. Such measures might be the replacement of a country's industrial equipment which might take 20 or 30 years or the replacement of automobile and lorry fleets by more economical models which would take 7-10 years.

The IEA describes the practice of effective conservation in its member countries ('Energy Conservation in the International Energy Agency 1978 Review') as an important precondition for continued economic growth. The IEA attributes the slowdown in energy consumption between 1973 and 1977 as compared with 1968 and 1973 (0.8% per annum as against 5.1% per annum) partly to economic recession, partly to higher energy prices, but also to the introduction of specific energy conservation measures in many countries.

Looking ahead the IEA saw significant potential for conservation in the transport sector especially for automobile fuel economy in North America. Potential savings in industry were estimated at 10-15% by 1985 and reductions in the energy needed to heat buildings at up to 40-50%.

Conservation is certainly viewed, in the industrialised world at least, as one way of reducing oil dependence. Much may be achieved by energy management - the coordination of the different methods of using less fuel to achieve the same result. For example, the installation of more efficient insulation to a building might be allied to a more sophisticated, less powerful heating system. Computerised heating controls can easily replace the wasteful practice of summer and winter thermostat settings with flexible responsive heating and cooling.

* i.e. below what consumption would otherwise have been

The Future

Heat pumps offer great potential for the heating or cooling of buildings - a heat pump absorbs heat at temperatures just below ambient then 'upgrades' it to a temperature well above ambient (although it can work equally well in the opposite direction, for instance in a refrigerator). The heat pump works through a vapour compression cycle but less energy is used to drive the compressor than is given out. Heat pumps are capable of 40% energy savings. About 2 million have been sold in the United States while in France it is illegal to advertise any other form of electric heating.

Conservation is not however, merely a matter of technology: to be effective it must be part of a universally adopted lifestyle and therefore both education in its justifications and its techniques as well as legislation to drive home the message are essential factors.

Reducing oil dependence must be the most immediate aim of conservation in industrialised countries. The reduction of oil used to produce heat may be accomplished by the switch to alternatives such as coal or natural gas. The same alternatives however, do not exist for oil the transport fuel where there is a huge potential for savings.

In the United States, cars account for about 31% of all oil consumed. If every driver, of the more than 100 million cars and 20 million light trucks in the US reduced their petrol consumption by just 5%, the savings would be the equivalent of some 17.9 million tonnes of gasoline per year.

If every household lowered its average heating temperatures by just six degrees, a further 28.4 million tonnes would be saved, while the raising of air conditioning temperatures by six degrees would produce savings of 9.5 million tonnes.

These three adjustments would mean savings of some 56 million tonnes of oil per year - almost the oil consumption of the whole of Africa and double the oil consumption of Sweden.

Conservation of transport fuel in the US is now being helped by the gradual increase of indigenously produced oils to world crude price levels. Unrealistic pricing, such as that in the US causes overconsumption, and, in the case of an oil importing country, much higher import bills than are necessary.

The reduction of oil import bills by means of conservation is recommended for developing countries as well as richer ones by the World Bank. In its report it estimates that the developing countries total energy demand of about 1,500 million tonnes in 1990 could be reduced by 220 million tonnes (15%) by strict 'energy demand management'.

In the LDC's, industry accounts for 35% of all the commercial energy consumed. Fourteen countries - including Brazil, Korea, India, Mexico and Romania use most of this by virtue of their involvement in energy intensive industries - steel, cement, ammonia, copper, aluminium, pulp and paper, and fertilisers. The World Bank considers that better recovery of by-products, and the installation of waste heat boilers and better insulation could, in conjunction with investments in retrofitting and attention to other energy saving techniques, produce a 15% energy saving by 1990 which would be worth about $19 billion and, in addition, would mean a 17% reduction in the cost of imports into LDC's. According to the World Bank the biggest reductions are possible in the pulp and paper industry (up to 25%) while 20% energy savings could be made in petroleum refining.

Transport, mainly road transport, accounts for 10-25% of LDC energy consumption. Here potential savings worth $11 billion in 1980 and $25 billion in 1990 are possible by cutting the oil import bill. The measures introduced would be the use of more efficient car, bus and lorry engines, and increased use of coastal shipping, rivers and railways for bulk transport.
transport.

Electricity accounts for 25% or so of LDC commercial energy consumption, a proportion which will increase to 30% by 1990 with power stations using $30 billion worth of oil. Here there is much scope for savings as much energy is wasted by random load shedding and inefficient transmission systems. In LDC's greater efficiency of operation could allow a 7% reduction in installed capacity by 1990 worth a $20 billion saving in capital investment and $2 billion a year in fuel bills. Switches from oil to coal fired power stations, the establishment of national grids plus the development of hydro electric and nuclear power would render further savings in oil imports possible.

Chapter 4

Transport

A quarter of the world's energy (excluding Communist countries) is used for transport. 90% of this energy comes from oil and the main usage sector is road transport. This means that over 20% of the world's oil consumption is burnt in car engines while in industrialised countries the proportion is far higher (31% in US). Increased world usage of motor cars is an inevitable consequence of the industrialisation of the LDC's. Present ownership levels in developing countries are very low compared to levels in industrialised countries. In Canada and the United States there is roughly a car to every two persons, in Japan six persons, in Brazil 16 persons, in Colombia 55, Indonesia 323, India 877 and Bangladesh 3,000.*

Reducing oil consumption in the heating and electricity generation sectors is straightforward in that other fuels may be substituted. With the use of oil as a transport fuel for cars, things are not so easy. In the face of inexorable increase in the demand for transport for the domestic user as a consequence of economic growth and population growth, three methods of reducing oil consumption present themselves. First and most radical a new means of propulsion other than the internal combustion engine might be considered. Secondly, new fuels for the internal combustion engine which do not involve the depletion of vital and limited energy resources. Thirdly, continuing to use oil fuelled internal combustion engines to drive cars but making the engines more efficient.

(A further alternative would be the abolition of personal private transport but since the flexibility offered by private transport is obviously much prized in all societies this alternative is not considered further. The energy saving advantages of rail transport, particularly when electrically powered, are, by virtue of numbers carried, obvious.)

The great advantage of an engine fuelled by petroleum is the high energy density of the fuel. Where particularly high energy fuels are a necessity - for example in aircraft - the only practicable fuel is kerosine (natural or synthetic). For surface transport, although an equally high energy density fuel is

* 1978 figures 'International Marketing Data and Statistics'

unnecessary an effective method of energy storage is vital. The following methods of energy storage have been tested - compressed air, fly wheels, heat energy in molten salts or hot refractory oxides and rechargeable batteries.

A number of these methods can be eliminated if conditions such as a minimum range requirement, ease of recharging and safety of operation are demanded. Compressed air does not have sufficient energy density for useful range: a flywheel would have to rotate at too high a speed for safety to ensure adequate range; storing heat energy is limited by the temperature of the working fluid and by losses due to conduction.

Rechargeable battery driven cars have attracted publicity for many years particularly since they are in regular use in businesses where a great deal of starting and stopping is necessary. Electricity driven vehicles fulfil all the basic requirements of the consumer except that, so far, a no battery offering a range comparable with that of a car has yet been made. Using lead/acid or nickel/iron cells a range of 100 kilometres between rechargings cannot be exceeded without using batteries with too higher weight as a ratio to total vehicle weight. However, even in the present stage of battery technology and without recourse to any of the many new battery types under test at present it is estimated that due to the preponderance of short trips* 70% of present fuel consumption in cars could be substituted by use of battery vehicles. One significant advantage of electric vehicles is that their widespread use would make far better use of off-peak electricity. This is because electricity generation is far more efficient when demand is consistent and the fact that recharging would take place at night plus the fact that peak transport demand is generally in summer, would mean that more generating plant could operate consistently at full load thus lowering the unit cost of electricity. A further advantage of electric vehicles is their lack of effect on the environment.

If we look far enough ahead electric vehicles are likely to replace the internal combustion engine but this is likely to be far in the future for a number of reasons. Firstly battery vehicles are much inferior in performance and range to internal combustion (IC) vehicles. Secondly the changeover would require such vast investment in new manufacturing and refuelling infrastructure that the incentive of a huge increase in fuel costs for IC vehicles would be necessary to render the switch economically viable. The use of electric vehicles this century will be very limited.

Turning to the use of fuels for IC vehicles which do not use up vital oil reserves or which may be used when oil is no longer available. The use of hydrogen has been suggested as one of the answers to future transport fuel

* This is in an industrialised country equivalent in size to the U.K.

needs. As a constituent of water, resources of hydrogen are virtually inexhaustible; the problem however is extracting it. Produced by electrolysis too much energy would be used up in the process to make it economic while the other method suggested by supporters of the 'hydrogen economy' - thermal dissociation of water from the heat output of a nuclear reactor is neither a tried nor tested technology.

The disadvantages of making methanol from the synthesis of hydrogen and carbon dioxide similarly lie in the energy cost of obtaining the two gases concerned, however alcohol from biological sources - biomass - has an application (see Synthetic Fuels section).

The growth of synfuel industries during the latter years of the twentieth century and the gradual increase of the contribution to oil demand made by these oil shale, tar sand, coal based fuels mean an extension of the life of the IC engine probably well into the mid-21st century.

Thus the IC car will certainly have private personal transport, still almost to itself in the year 2000 but changes will be seen in the nature of the car - in its size, efficiency and economy. Changes in aerodynamics, lighter alloys in engines and leaner fuel consumption will be the main features. In the area of economy the microprocessor will have a growing effect, enabling engines to be programmed for optimum performance under different driving conditions.

There will also be changes to accommodate to the use of synthetic fuels. Car fuel systems and carburettors must be redesigned for alcohol driven cars, which require fuel tanks double the size of petrol driven cars.

Efficiency improvements in cars are likely to be responsible for significant conservation of fuel in the next 10 years. In Western Europe new cars are estimated to become on average 2% more fuel efficient per year to 1990, Europe's motor manufacturers having given the commitment to Governments to improve the fuel efficiency of their 1985 models by 10% over the 1978 efficiency level.

In addition to improvements in fuel consumption by means of improved design, weight reduction and increased engine and transmission efficiency fuel is likely to be conserved by the more efficient use of vehicles. This may be achieved through better driving habits, improved traffic flow systems and road networks and car pooling.

Chart 28

USERSHIP OF ENERGY IN 1979

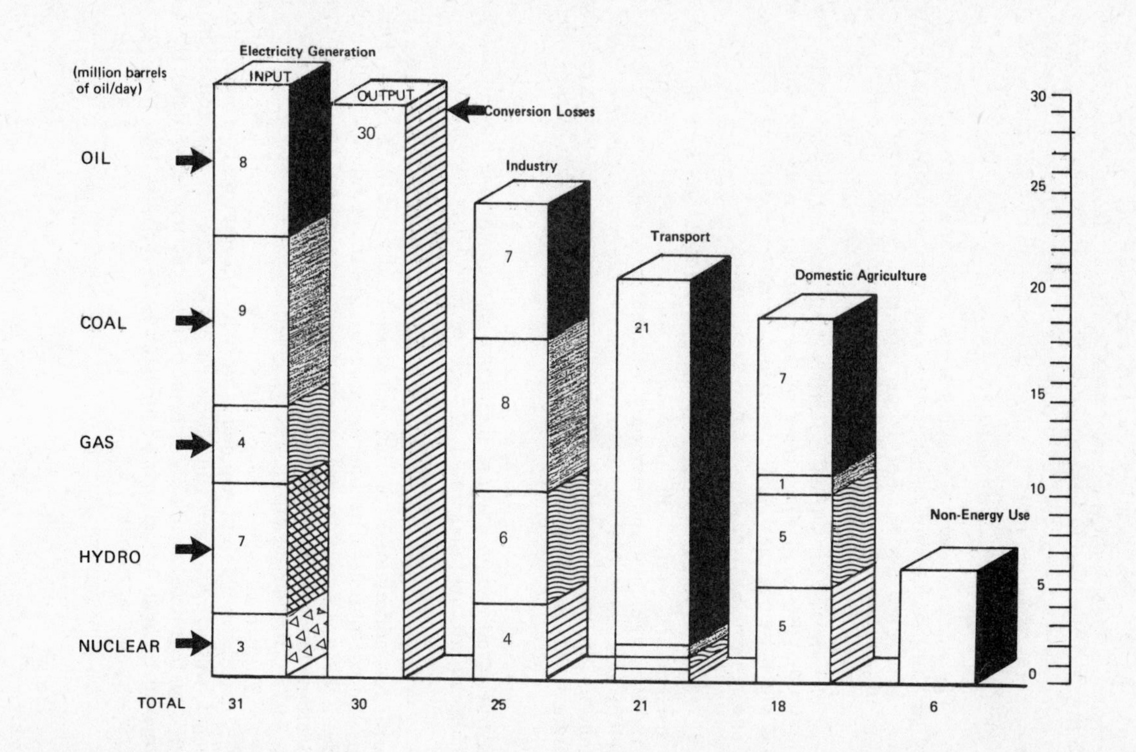

Chapter 5

Energy - How We Use It

The essence of any energy system is the conversion of energy from one form to another. By means of photosynthesis,plants convert the radiant energy of the sun into chemical energy. Thus the burning of coal, formed as it is of vegetable matter, is a stage in the conversion of solar energy. The First Law of Thermodynamics basically states that energy has never been observed to be created or destroyed. Thus when one energy form is converted to another output is the same as input. However, some forms of energy are more useful than others. The heat energy within a kettle of boiling water is not lost when that kettle is poured into a bath of cold water but the water in the bath is not hot enough to make tea (it is 'low grade heat'). The energy, once transferred to the bath is in the form of low grade heat and therefore virtually useless. Conversion efficiency as a percentage is therefore usually expressed as:

$$\frac{\text{useful energy output}}{\text{total energy input}} \times 100$$

or the proportion of the useful energy produced to the energy content of the fuel that produced it.

The conversion efficiency of the process of electricity generation using fossil fuels is between 30% and 40% when at its most efficient. This means that for every 100 units of energy in the coal which heats the boiler that makes the steam that drives the turbo generator only 30 units of energy in the form of electricity are produced.

Making electricity is, in fact, the main use of the primary fuels - accounting for 31% of the world's total energy input - the main fuels used being coal (29%), oil (26%), hydro (22%), gas (13%) and nuclear (10%). By the year 2000 coal and nuclear will be the two leading fuels for electricity generation, accounting together for about 65% of the total.

But electricity generation is only a stage in the energy chain; the electricity is then used in industry and in the home for lighting, heating and powering machinery. The electricity generated, however, due to the amount of heat lost in the process of generation (mainly at the stage of converting heat energy of steam into the mechanical energy of the turbine) only represents about 30% of the energy used in the power station. Even more is lost during the transmission process; the lower the voltage the more energy is dissipated.

Combined heat and power schemes attempt to use the heat wasted in the process of electricity generation but this is a difficult process because the heat is low grade heat.

Despite the considerable losses of useful energy in the conversion process, electricity is indispensable for its flexibility of use.

The main final use of energy is in the industrial sector where 37% of all energy use occurs, while transport accounts for 33% and the domestic/agricultural sector 30%.

The difficulties involved in reducing the world's oil dependence are clearly seen in the fact that the transport sector is almost completely oil fuelled. The large percentage of oil used in power stations is more easily reduced by substituting alternatives.

Chapter 6

Beyond the Year 2000

The Renewables

A 'renewable energy source' is one which does not depend on finite reserves of either fossil or nuclear fuels. The energy upon which the various renewable sources draw is the sun, in the case of solar energy, wind energy, wave energy, biomass and hydro electric power, while in the case of tidal energy the source is the moon, and the earth's rotation. All of these renewable sources but solar power (when used directly) and biomass involve the generation of electricity. Biomass - energy derived from vegetable matter - is the only renewable with chemical application.

Although the energy source is not in the strict sense infinite, geothermal and fusion power are also generally defined as renewable sources since there is no prospect of either source running out.

The only one of the renewable sources currently commercially exploited on a significant scale is hydro electric power which in 1979 contributed about 6% to world energy supply. Hydro electric power will still be the main source of renewable energy in the year 2000 and will also be responsible for most of the growth in usage of renewable sources mainly by virtue of projects in South America. South America has the largest hydro electric potential due to the potential energy contained in its rivers by virtue of their altitude and volume. The largest hydro electric scheme in the world is at Itaipu on the river Parana in Brazil (see South America section).

Most of the Western World's best sites have now been developed but potential still exists in less developed countries; Zaire, for example, contains half of Africa's hydro electric potential.

The advantages of hydro electric power are its high conversion efficiency - more than 80% as compared with efficiency rates of 30% or so in thermal power stations. Also HEP schemes generally involve dams which may be useful as flood barriers and for irrigation schemes, particularly in developing countries. The main disadvantages are the cost of HEP projects and the fact that generation is dependent upon volume of water available, thus other generating capacity must always be available to cope with times of drought.

Hydro electric power, though renewable, has a finite limit to its development potential unlike some of the other renewables and will never contribute much more to world energy supply than it does at present.

The renewable energy sources must eventually dominate world energy supply. The speed with which this will occur depends primarily upon the speed with which the prices of conventional fuels rise. Prices will be decided by such factors as the supply demand relationship, the cost of recovery as the search for fossil fuels moves to less and less hospitable regions and as the shift to more highly processed secondary fuels such as substitute natural gas and synthetic liquid continues.

Expenditure on renewable energy sources in Europe is, with the exception of fusion power very low. The leading EEC country in expenditure over the five years 1974-1978 was France followed by Germany and Italy with U.K.,a poor fourth.

The Decision to Change

Each time crude oil prices take an upward leap a proportion of world energy consumption turns away from one energy supply mode to another. That is to say, a decision is made to turn from one mode to another. When such decisions are taken there is a time lag due to the lead time involved in bringing the new production mode on stream. Lengths of lead times add considerably to initial costings and such factors must be balanced carefully in the decision making process, along with size of investment necessary and energy production costs.

Renewable sources of energy generally require larger investment than conventional energy except in the case of high temperature geothermal and some types of bio conversion. Production costs for the renewables and, even more so, for synthetics are far higher than for coal, oil or gas. Despite the sophisticated techniques involved in enhanced recovery of oil and recovery of heavy oil both investment and production costs are lower than for tar sands, oil shale and for the renewables. Nuclear power has long lead times (particularly in the US) and requires a higher investment than many synthetics. However its energy production costs are lower.

Chart 29 THE RISING COST OF INVESTMENT

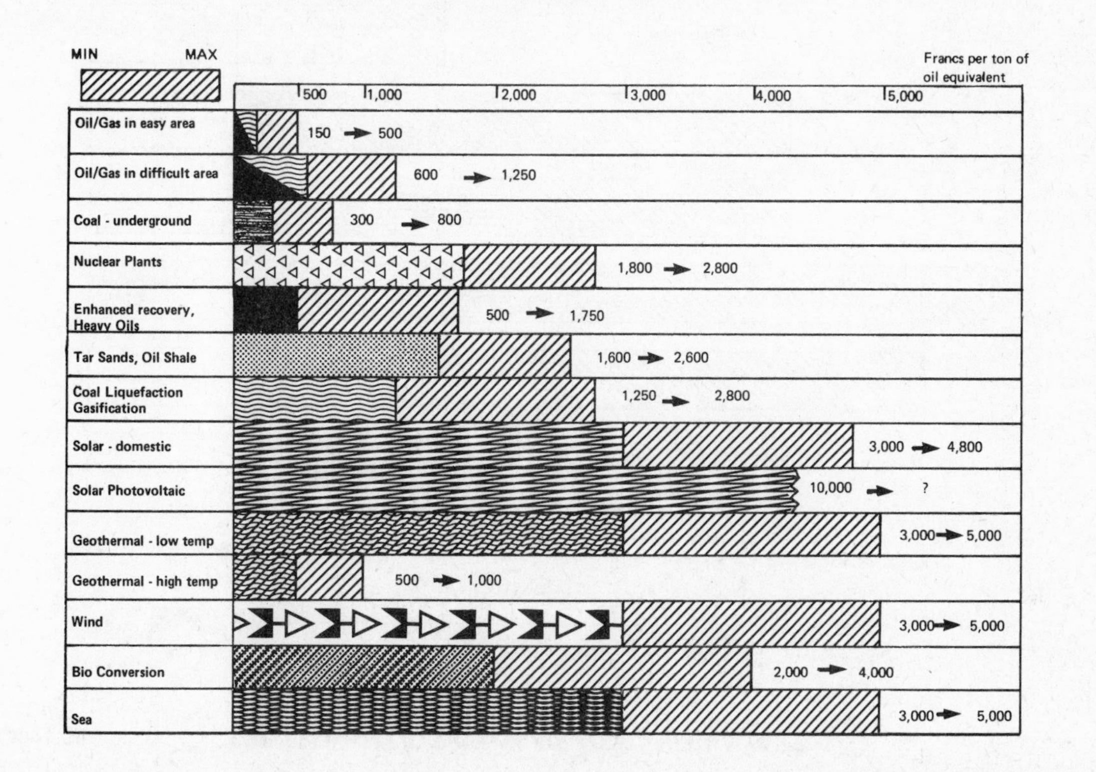

Chart 30

THE PRODUCTION COSTS OF NEW ENERGY SUPPLY *

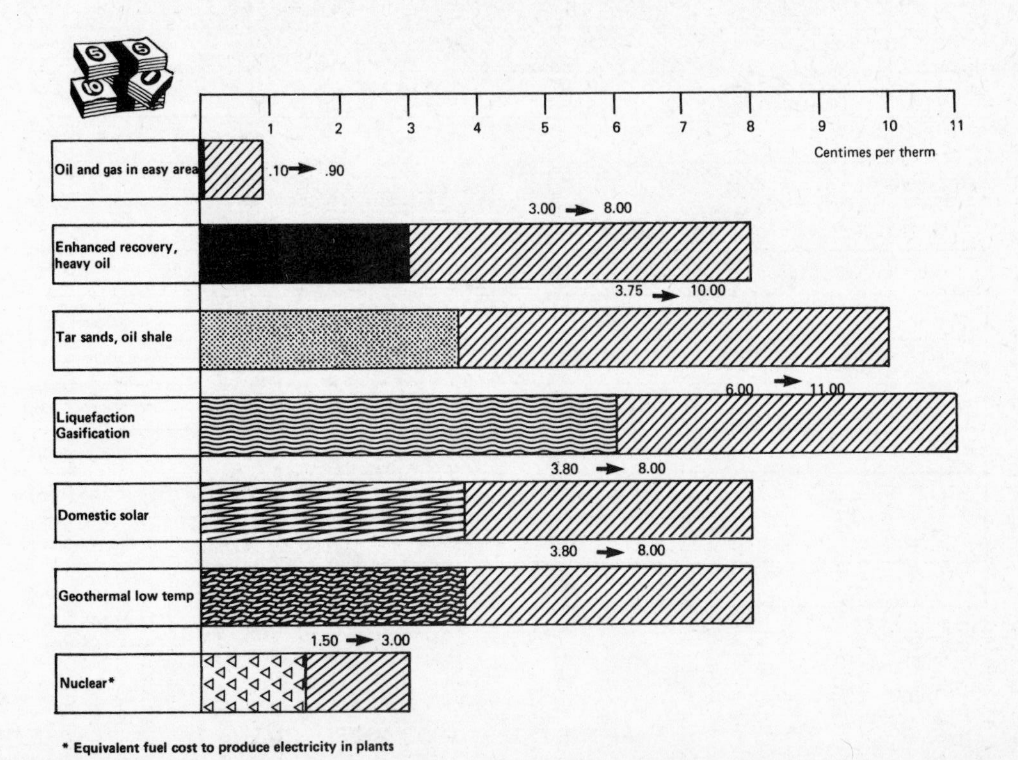

Wind Power

The power of the winds, worldwide, may be estimated at three or four times total world energy consumption. However, it will never be practicable to convert more than a small percentage of this into useful energy. However, a project is under consideration in the US for some 300,000 windmills spread across the plains which would supply some 200,000 MW more than half the present US installed capacity. For most parts of the world, however, the most convenient and effective sites for wind generators would tend to be coastal or offshore where winds are stronger and space is not at a premium.

The major factors governing the use of wind generators are as follows. Power available from the wind obeys the 'rate of the wind speed' by which a doubling of wind speed increases the power available by a factor of eight. This means that higher wind speeds must be sought - one reason for the choice of coastal and offshore sites.

The second point is that power available is proportionate to the square of the propeller diameter.

Thirdly, output increases with the height of the blades since, lower down, wind speed is reduced by friction.

The largest wind generator in operation is probably the NASA MOD 1 at Boone, in North Carolina (sponsored by the US Department of Energy) which has a 200 foot diameter and is capable of generating two megawatts of peak power. Modern computer technology renders generation more efficient; a mircoprocessor predicts wind behaviour and adjusts the wind generator to maintain maximum performance.

One advantage of wind power is that it would tend to generate best when weather is bad which is also the time of peak need for electricity in many countries. However, as with other renewables the lack of control over the matching of electrical output to demand is a disadvantage.

Increased usage of renewable energy sources, most of which will mainly convert energy into electricity, will necessitate better methods of storage of electrical energy than at present. Possibilities include the use of surplus electricity to compress oil into huge reservoirs, the electrolysis of water into hydrogen and oxygen and pumped storage (see Electricity section).

According to a Swedish study of the economics of wind power in 1979 recommending the construction of 700 offshore windmills, the total cost of generation by this method was comparable with that of AGR generation at a ratio of 1 (nuclear) to 1.2 (wind power). Although capital and operating costs of the wind generators were higher, the disparity was reduced by the absence of fuel costs. (Estimated nuclear 1.70 p/Kwh/wind 2.00 p/Kwh).

A large aerogenerator programme is at present underway in the Netherlands where it is hoped that wind will be supplying about one quarter of the Dutch electricity demand by the year 2000. However, this will necessitate some 5,000 aerogenerators and strong environmental objections are expected.

As the renewable energy sources develop they will inevitably quickly lose their present quaint, pastoral image and the concepts of noise and horizon pollution will assume as significant a role in the public bestiary as do atmospheric and nuclear pollution today. The level of noise pollution from one of the larger aerogenerators which during high winds would achieve supersonic speeds at the end of the blade would certainly be of formidable proportions.

Where climatic and environmental factors make land based aerogenerators impossible, costs would be high. According to an assessment commissioned by the U.K. Department of Energy the cost of the support structure alone for one wind turbine (excluding the turbine) would be £3 million while one machine would cost £34,000 per year to maintain. The support facilities on shore for a 200 machine 600 MW cluster of wind generators would be around £25 million while transmission costs were estimated at £171-189 kilowatts - four times that for a conventional 600 MW power station (all mid-1979 prices).

It has been estimated that the capital cost of a 1,000 MW installation on the UK coast could be comparable with the cost of a conventional power station of the same capacity but this does not take into account the formidable amount of R and D necessary both to develop a 2 MW, 100 metre diameter aerogenerator and to design the kind of structures and transmission systems to withstand the rigours of the ocean.

Power from the Sea

Wave energy conversion offers huge amounts of power but demands a great deal more new technology than wind energy. Wave power offers a higher energy reward than tidal power - the sea acts as a gigantic energy storage battery for the energy of the wind.

In the U.K., where a comparatively large amount of research has been carried out into wave power it is estimated that with wave generators along the 1,450 kilometre coastline 50% of the U.K.'s electricity requirements could be supplied. (The power flow from Western Atlantic waves is around 70 kilowatts per metre of wave front.)

The problem is to develop a device which as well as demonstrating a high degree of efficiency in converting wave energy into electricity, is also robust enough to stand up to the buffeting and corrosion of the sea.

The Salter's Duck, designed by Stephen Salter converts the bobbing motion from which the name derives into the rotation required to drive a generator. The Cockerell Raft, invented by Sir Christopher Cockerell undulates with the waves and energy is converted via hydraulic compressors, to the generators. In 1978 it was claimed that these devices could not produce electricity at a cheaper rate than 20 pence per kilowatt hour (as opposed to the Central Electricity Generating Board's 2.7 pence per kilowatt hour) but this has now been reduced to 10 pence per kilowatt hour while Dr. Salter maintains that the Duck could achieve 4 pence per kilowatt hour.

A further promising device is the Kaimei which uses the changes in air pressure in a tube caused by the rise and fall of water, to drive an air turbine.

Any of these devices could obviously offer more attractive price incentives were they mass produced but at present fairly small amounts are being spent on wave energy research, which has, so far, not caught the imagination of large investors whether Government or private, in the same way as, for instance, nuclear fusion. This may change following the results of the full scale trials of one system off the North West coast of Scotland, but prior to this a decision must be made as to which of the systems on offer, looks to have the best potential.

There are also other ways of deriving energy from the seas. OTEC stands for Ocean Thermal Energy Conversion. This system exploits the temperature differential between different depths. A low boiling point liquid such as ammonia is vaporised by the warm surface water. The expanding vapour drives a turbo generator while the ammonia is reliquefied in a condenser by cold water pumped from as deep as 3,000 feet in a self contained operation. A number of demonstration OTEC plants are operating at present.

The OTEC system, although low in output, has the advantage of generating at a fairly constant level. Wave conversion devices will vary in output according to the power of the waves and where effective storage of energy is impractical, wastage must ensue.

Tidal schemes have the advantage of regularity, an integrated electricity system could organise itself according to high and low water. Tidal power

Chart 31 WAVE ENERGY DEVICES

SALTER'S DUCKS

COCKERELL'S RAFTS

(Artist's impressions of wave energy devices. The human figures are to scale.)

Source : ATOM

systems operate by filling a single basin or river estuary at rising tide through sluice gates then emptying it during falling tide through turbines. There are, in fact only a limited number of sites in the world with sufficient tide range to make such schemes cost effective. The largest tidal ranges are found in areas such as the Bay of Fundy (16 metres), the U.K.'s Severn Estuary (13-14 metres), the Rance Estuary in France (12 metres) and sites in the White Sea and the Sea of Okhotsk in the USSR, the Yellow Sea in Korea and the Kambay and Kutch bays in India. The only commercial size plant in operation at present is at La Rance. This has a capacity of 240 MW, between a quarter and one eighth the size of the capacity of a large power station.

Schemes are, however, being considered at the Bay of Fundy, which has 37 sites suitable for tidal power schemes, and the Severn in the U.K., although the economic viability of the latter is still uncertain.

Solar Power

In one way or another almost all the world's energy resources derive from the sun. The sun directs 100,000 BTOE worth of energy at the earth per year, a quarter is reflected back and 75,000 BTOE is absorbed but only 30 BTOE is stored. This 30 BTOE is transformed by photosynthesis into the chemical energy of plants. Over the 400 million years which vegetation has existed 7,600 BTOE of energy from plants has been converted into fossil fuels (at a rate of about 16,000 tonnes per year). The part of the sun's energy absorbed by the earth's atmosphere creates the winds and the waves on the sea.

Of the 75,000 BTOE of solar energy which reaches the earth, only 2.3 BTOE is converted to useful energy by man - 71% is used for food production, 24% goes on the burning of wood and agricultural waste and 4% on hydro electricity. The direct harnessing of the sun's energy - solar power - is along with nuclear fusion the renewable energy sector which has attracted the most interest.

In less than one week the earth receives as much solar energy as the world's total fossil fuel reserves. Solar energy is produced by the thermonuclear reactions which are constantly taking place in the sun which has a temperature of about 20 million degrees centigrade.

There are many ways in which solar energy may be directly exploited by man. The main disadvantage to direct use lies in the fact that supply reaches its maximum at the same time as demand is at its minimum.

The simplest method of conversion is by flat plate collectors in which a fluid, usually water, may be heated.

Focussing collectors are more sophisticated devises which produce more concentrated heat. The Odeillo solar furnace in the Pyrenees achieves a temperature of 4,000^{o}C while larger furnaces are being developed, again at Odeillo and in Sicily, capable of a 1 MW capacity.

For these types of convertors heat conversion is direct and storage is difficult. When solar power can be converted directly into electricity it becomes much more usable. The photovoltaic cell is, at present, the object of a great deal of research and investment but progress must yet be made in the chemistry of certain materials, particularly silicium, to bring down the cost fiftyfold before this method becomes economical.

A further area of solar energy research uses the difference in potential between the seams of two metals brought to different temperatures. These are what are known as heat batteries or thermo-ionic cells. However these also will require many years of development.

The most ambitious scheme of all for the harnessing of solar energy is the SPS solar power satellite - an idea devised (and patented) by Dr. Peter Glaser. The whole scheme would involve the placing of a 'solar sail' 10 kilometres by 5 kilometres in size, in geostationary orbit 23,000 miles out from the earth where the sun would never be out of sight. The power of the sun would then be harvested by an array of silicon or gallium arsenide solar cells which would then be beamed back to earth by microwave and converted into low-frequency, alternating current electricity and fed into the USA's grid. The electricity output of one SPS would be 5,000 MW, bigger than any single power station in the world. The cost of one SPS according to the US National Acronautics and Space Administration was in 1979 put at $25 billion while $30 billion would be required to set up the project. Should the systems have been accepted it was considered that the first commercial power satellite could have been in operation by 2010.

The SPS will probably never be attempted but its theoretical viability is suggested by the fact that NASA took three years and spent $16 million on assessing the concept.

In weighing the cost of the venture it should be remembered that 2.5 million people were employed and $25 billion spent on the project to land a man on the moon in the 1960's.

Less grandiose is the scaling up of the solar furnace idea. In a device called the heliostat or 'power tower' gigantic mirrors are focussed upon a central, elevated boiler. However, to generate 100 MW of electricity - one tenth the capacity of a large power station - the sunlight falling on 3-5 square kilometres would have to be caught.

The technology for the effective use of solar power to produce heat exists today and, could, in the right climatic conditions, save a great deal of fossil fuel with widespread use, either to produce low temperature or, when focussed, high temperature industrial or domestic heating. However, the economics of such systems are not viewed with very great enthusiasm. The use of sunlight to generate electricity is being vigorously pursued - a great deal of money is being invested in the improvement of photovoltaic cell technology - but looking to the future some significant studies in the storage of electricity must be made before solar generation - or any of the other renewable techniques which cannot produce power on demand - will really come into their own.

Biomass

The immense amount of solar energy which growing plants capture by photosynthesis and convert into chemical energy - some 30 billion tonnes of oil equivalent per year or more than four times the world's yearly total commerical energy consumption - is a renewable source of energy which offers great potential.

The proportion of this energy conversion accounted for by agriculture, however, is small, but even this is not exploited.

The most obvious exploitation of biomass is the manufacture of alcohol by fermenting surplus or specially grown crops - e.g. sugar, corn, bananas (see South America and Synthetics sections). This is particularly important since biomass is the only renewable energy source which does not offer energy either as direct heat or electricity.

Merely considering biomass as the exploitation of waste shows great unused potential. A study by the U.K. Department of Energy's Technology Support Unit (ETSU) at Harwell revealed that about 60 million tonnes coal equivalent of energy - 18% of total U.K. energy consumption - was thrown away every year as waste. This was made up of the organic residues of farms, forestry, industry and domestic refuse.

Many countries of the world could, not only make better use of wastes, but increase agricultural production of the crops of which surplus produce could best be used to manufacture fuels.

Although alcohol is the most obvious example of a biomass fuel others would include methane from sewage which could be used in the internal

Chart 32 **GEOTHERMAL POWER - HOT ROCKS TECHNIQUE**

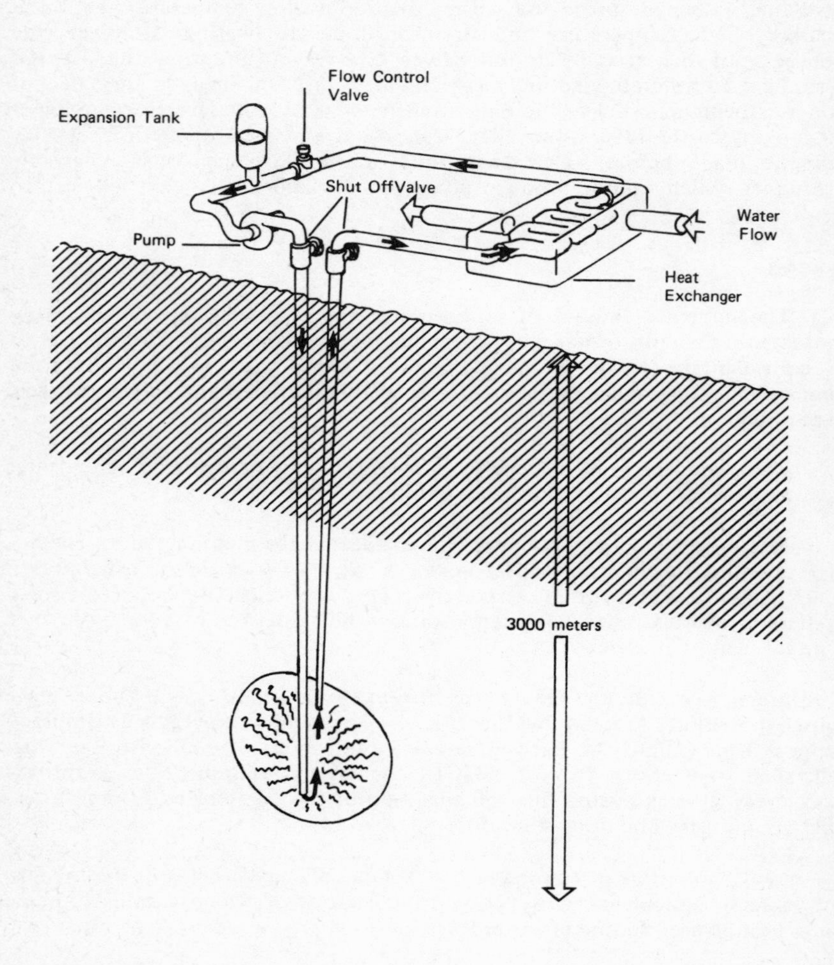

Source : ATOM

combustion engine as well as being a chemical feedstock.

Brazil and the United States are the main exploiters of biomass energy but Canada, China, India, the Philippines, Australia, New Zealand and South Korea all have demonstration or pilot projects.

Geothermal Power

Geothermal energy is based on the increase in underground temperature which is, on average, 1°C per 30 metres of depth but in certain areas faults at plate boundaries allow higher temperatures at shallower depths.

There are basically three methods of recovering geothermal energy. The first is to drill for and locate hot water (having first mapped the geothermal gradient of the ground since conditions vary considerably). A large number of wells are then sunk into the hot water deposits, thus allowing circulation. The water rises, the heat is transferred to heat exchanges and the water then returns below ground to be reheated. This system may exploit water at temperatures of between 60°C and 12°C and is widely used today in France, USSR, Hungary, Iceland and many other locations. One of the best known sites is at Villeneuve-la-Garenne, Paris, which has been converted from oil to geothermal heating, with an annual saving of 3,400 tonnes of oil. In Reykjavik, Iceland, geothermal water provides almost all heating.

The second method of using geothermal energy makes use of high temperature steam - in essence an attempt to domesticate the geyser - to drive turbo generators and produce electricity. One such plant has operated at Laredello near Florence since 1905 while the largest present application is at Geysers Falls, California. However, such sites are extremely rare.

The hot rocks technique holds out the best hope for exploitation of geothermal energy but the technology has not yet been perfected. The problem with hot water is that the heat is low grade and in addition suitable sites must be explored for like oil.

The idea of drilling down to where rocks are at high temperatures, excavating a large cavity then pumping down a fluid which could be vaporised and used to generate electricity could give geothermal energy a wide application but the techniques involve the development of new technology. An experiment to establish the feasibility of this technique is now in progress at Los Alamos in New Mexico, where suitable rocks are within 2,000-3,000 metres of the surface.

Although, on a world scale, the geothermal energy contribution is low, more than 50 countries are, at present engaged in research into methods of exploiting it.

Nuclear Fusion

Nuclear fusion - the process which powers the sun is the Holy Grail of energy production. The fuels which the fusion reaction would use are deuterium (a common isotope of hydrogen) and lithium and the energy potential exists to supply the world with energy for at least 5,000 years.

The large scale release of fusion energy has so far occurred only in stars and in the hydrogen bomb.

In the fusion reaction atoms of hydrogen would be fused to make helium and, in the process, a huge amount of energy is released. Controlled fusion has fewer obvious negative aspects than fission. There are no serious long term spent-fuel reprocessing or waste problems nor is there a potential for the large scale release of radioactivity.

However, the controlled fusion reaction has not yet been achieved experimentally. The difficulty lies in heating the fuel to temperatures of about 100 m^{o}C and in containing any substance heated to such a temperature. At this temperature atoms form a 'plasma' which carries an electric charge and can thus be kept away from the walls of the container by a magnetic field. So far, experiments in Russia and the USA have not managed to create a reaction in which the energy output exceeded the energy input.

There are two types of vessel under consideration for the containment of the controlled fusion reaction - the doughnut shaped tokomak or torus and a cylinder shaped container. In the past the torus has been the more in favour as its plasma containment is more efficient. However, the cylinder shape is easier to heat (a machine at Livermore, California has achieved 150m^{o}C).

At present the US has a $335 million annual budget for fusion research (magnetic). A $226 million fusion furnace is to be built at Livermore while a $294 million device is nearing completion at Princeton Plasma Physics Laboratory in New Jersey. Neither of these are intended to achieve more than the 'break-even point' at which energy input to heat the fuel is equalled by output from the fusion reaction.

In Britain a project called JET (Joint European Torus) and costing £130 million, to build a European tokamak is underway. The project is funded jointly by the EEC and Sweden and commissioning is likely to be in 1983.

Research is also in progress to explore the possibility of creating fusion by a technique of laser bombardment but this technology is at a far earlier stage than 'magnetic fusion' which itself is unlikely to achieve commercial application within the next 40 or 50 years, if ever.

Part Five
STATISTICAL TABLES

Table 1. **WORLD PRIMARY ENERGY CONSUMPTION 1969-1979**

(million tonnes of oil equivalent)

	MTOE	% Increase Over Previous Year	Index (1969=100)
1969	4,900.8	-	100
1970	5,178.8	5.7	106
1971	5,397.0	4.2	110
1972	5,637.7	4.5	115
1973	5,926.7	5.1	121
1974	5,970.5	0.7	122
1975	5,983.9	0.2	122
1976	6,318.1	5.6	129
1977	6,513.7	3.1	133
1978	6,743.5	3.5	138
1979	6,960.4	3.2	142
1990	9,506.0	-	194
2000	12,296.0	-	251

(Yearly Change: 1979 over 1969 + 3.6%; 1979 over 1974 + 3.1%)

Source: BP / Own Calculations.

Table 2. WORLD PRIMARY ENERGY CONSUMPTION BY PRIMARY FUEL 1969-1979
(Unit: million tonnes oil equivalent)

	OIL			COAL			NATURAL GAS			Total Energy
	MTOE	% of Total	Index (1969 = 100)	MTOE	% of Total	Index (1969 = 100)	MTOE	% of Total	Index (1969 = 100)	MTOE
1969	2,099.8	42.8	100	1,614.6	32.9	100	877.5	17.9	100	4,900.8
1970	2,284.0	44.1	108	1,640.9	31.7	102	928.7	17.9	106	5,178.8
1971	2,417.1	44.8	114	1,636.7	30.3	101	997.6	17.9	114	5,397.0
1972	2,389.4	45.9	121	1,638.2	29.0	101	1,048.4	18.6	119	5,637.7
1973	2,793.4	47.1	131	1,673.4	28.2	104	1,082.0	18.3	123	5,926.7
1974	2,756.9	46.2	110	1,694.7	28.4	105	1,111.7	18.6	127	5,970.5
1975	2,722.8	45.5	122	1,714.5	28.6	106	1,104.6	18.4	126	5,983.9
1976	2,897.3	45.9	131	1,790.2	28.3	111	1,166.0	18.4	133	6,318.1
1977	2,985.6	45.8	134	1,836.8	28.2	114	1,187.4	18.2	135	6,513.7
1978	3,083.2	45.7	138	1,879.3	27.9	116	1,231.4	18.3	140	6,743.5
1979	3,119.6	44.8	139	1,976.6	28.4	122	1,296.6	18.6	148	6,960.4
Yearly change 1979 over 1969		3.3			2.0			4.0		
Yearly change 1979 over 1974		1.9			3.1			3.1		
1990	3,571.0	3.8	170	3,091.0	3.2	191	1,799.0	1.9	205	9,506.0
2000	4,109.0	3.3	196	4,543.0	3.7	281	2,081.0	1.7	237	1,229.6

Source: BP / Own Calculations

Table 3. **WORLD ENERGY CONSUMPTION – MAIN SOURCES OTHER THAN FOSSIL FUELS 1969-2000**

Unit: Million tonnes of oil equivalent

	WATER POWER			NUCLEAR			Total Energy
	MTOE	% of Total	Index (1969 =100)	MTOE	% of Total	Index (1969 =100)	MTOE
1969	292.9	6	100	16.0	-	100	4,900.8
1970	305.4		104	19.8		124	5,178.8
1971	317.6		108	28.0		175	5,397.7
1972	323.3		110	38.4		240	5,637.7
1973	328.5		112	49.4		309	5,926.7
1974	344.6	6	118	62.6	1	391	5,970.5
1975	354.9		121	87.1		544	5,983.9
1976	358.2		122	106.4		665	6,318.1
1977	371.9		127	132.0		825	6,513.7
1978	399.8		136	149.8		936	6,743.5
1979	411.8	6	140	155.8	2	974	6,960.4
Yearly Change 1979 over 1969	3.5				25.6		
Yearly Change 1979 over 1974	3.6				20.0		
1990e	529.3	6	181	506	5	-	9,506.0
2000e	681.5	6	233	882	7	-	12,296.0

Source: BP / Own Calculations.

Table 4. **WORLD ENERGY CONSUMPTION BY PRIMARY FUEL 1970–2000**

(Unit: Million tonnes of oil equivalent

	OIL		COAL		NATURAL GAS		HYDRO*		NUCLEAR		TOTAL	
	MTOE	% of total	MTOE	% of total	MTOE	% of total	MTOE	% of total	MTOE	% of total	MTOE	% of total
1970	2,284.0	44	1,640.9	32	928.7	18	305.4	6	19.8	-	5,178.8	100
1975	2,722.8	45	1,714.5	29	1,104.6	18	354.9	6	87.1	1	5,983.9	100
1979	3,119.6	45	1,976.6	28	1,296.6	19	411.8	6	155.8	2	6,960.4	100
1990	3,545	37	3,091	33	1,799	19	529.3	6	506	5	9,506	100
2000	4,109	33	4,543	37	2,081	17	681.5	6	882	7	12,296	100
% change 1970–79	+37		+20		+40		+35		+687		+34	
% change 1979–2000	+32		+130		+60		+65		+466		+75	

* and renewables

Source: BP / Own Calculations.

Table 5. **ENERGY CONSUMPTION – MAIN CONSUMERS 1979–2000**

	1979		1990		2000	
	MTOE	% of total	MTOE	% of total	MTOE	% of total
USA	1,148.0	27.3	2,330	24.6	2,949	24.0
USSR	1,148.0	16.5	1,638	17.3	1,978	16.1
China	576.9	8.3	915	9.6	1,201	9.8
Canada	223.4	3.2	297	3.1	357	2.9
W. Europe	1,325.6	19.0	1,644	17.3	1,867	15.2
Japan	380.7	5.5	502	5.3	640	5.2
Middle East	105.8	1.5	218	2.3	485	3.9
Africa	143.1	2.1	207	3.2	608	4.9
Australasia	84.8	1.2	113	1.2	148	1.2
Latin America	314.4	4.6	486	5.1	674	5.5
E. Europe	441.8	6.3	572	6.0	707	5.7
S. Asia	129.3	1.9	195	2.0	285	2.3
S.E. Asia	183.5	2.6	289	3.0	395	3.2
World	6,960.4	100	9,506	100	12,296	100
Communist World*	2,166.0	31.0	3,125	32.8	3,886	31.1
Non-Communist World	4,794.0	68.9	6,381	67.1	8,410	68.9

*Not including Yugoslavia, Cuba.

Source: Own Calculations.

Table 6.

THE WORLD'S MAJOR ENERGY CONSUMERS
CONSUMPTION OF ENERGY – 1969–1979

(Million tonnes of oil equivalent)

	1969	1972	1975	1978	1979	Yearly % Change	
						1979 over 1969	1979 over 1974
USA	1,626.5	1,767.8	1,722.1	1,903.5	1,898.1	+1.6	+1.4
USSR	712.8	836.5	909.8	1,104.9	1,148.0	+4.9	+4.4
China	255.7	336.1	420.2	505.7	576.9	+8.5	+7.9
Japan	353.1	317.4	339.9	363.6	380.7	+4.2	+1.5
W. Germany	229.9	241.7	242.6	270.0	285.0	+2.5	+2.2
Canada	152.7	179.7	195.0	217.9	223.4	+3.9	+2.5
U.K.	208.2	215.6	204.3	211.2	220.8	+0.6	+0.6
France	146.4	171.4	171.1	191.8	195.4	+2.9	+1.2
Italy	110.5	132.5	132.9	145.4	147.8	+3.0	+1.5
Netherlands	53.1	72.3	71.3	74.1	75.7	+3.6	+1.2
Spain	43.3	55.0	64.4	75.1	77.2	+6.0	+4.0
Belgium/Luxembourg	45.3	51.3	46.7	51.4	52.7	+1.5	+1.0
Sweden	38.9	44.0	45.1	42.2	44.3	+1.3	+0.6
Norway	23.3	26.9	28.8	29.2	31.3	+3.0	+2.0
Switzerland	19.3	21.5	23.6	24.9	24.4	+2.4	+1.6
Yugoslavia	19.7	26.7	32.0	35.8	37.9	+6.8	+4.2
Austria	18.7	22.1	23.4	25.0	26.1	+3.5	+2.2
Turkey	12.2	19.4	19.7	23.1	23.1	+6.6	+5.6

Table 6. continued

	1969	1972	1975	1978	1979	1979 over 1969	1979 over 1974
Denmark	18.7	20.7	18.2	19.5	20.3	+0.9	+2.3
Finland	14.5	16.3	17.4	20.0	21.6	+4.1	+4.7
Greece	9.0	13.4	17.0	19.8	20.4	+8.6	+6.5
Eire	6.8	5.9	6.1	6.8	7.0	+0.4	+2.5
Middle East	59.1	78.7	94.6	114.4	105.8	+6.0	+1.9
N. America	1,779.2	1,947.8	1,917.1	2,121.4	2,121.5	+1.8	+1.5
W. Europe	1,019.0	1,170.1	1,176.5	1,279.4	1,325.6	+2.7	+1.8
E. Europe	308.8	353.9	387.7	434.8	441.8	+3.6	+3.8
Latin America	1887.	223.1	261.7	306.2	319.4	+5.4	+4.6
S.E. Asia	90.3	115.3	133.0	171.0	183.5	+7.3	+7.3
Africa	86.3	96.0	106.2	135.0	143.1	+5.2	+7.1
S. Asia	93.1	99.2	105.7	125.3	129.3	+3.4	+4.5
Australasia	54.7	63.9	71.5	82.1	84.8	+4.4	+3.9
Eastern Hemisphere	2,932.9	3,467.1	3,805.1	4,315.9	4,519.5	+4.4	+3.8
Western Hemisphere	1,967.9	2,170.6	2,178.8	2,427.6	2,640.9	+2.2	+1.9
World (excl. Communist Bloc)	3,623.5	4,111.2	4,206.2	4,698.4	4,793.7	+2.8	+2.3
World	4,900.8	5,637.7	5,983.9	6,743.5	6,960.4	+3.6	+3.1

Source: BP

Table 7. **THE WORLD'S MAJOR ENERGY CONSUMERS**
CONSUMPTION OF ENERGY 1969–1979

(Index 1969 = 100)

	% of Total	1969	1972	1975	1978	1979	% of Total
USA	33.2	100	109	106	117	117	27.3
USSR	14.5	100	117	128	155	161	16.5
China	5.2	100	131	164	198	226	8.3
Japan	5.2	100	125	134	144	150	5.5
W. Germany	4.7	100	108	106	117	124	4.1
Canada	3.1	100	118	128	143	146	3.2
UK	4.2	100	103	98	101	106	3.2
France	3.0	100	117	117	131	133	2.8
Italy	2.2	100	120	120	131	134	2.1
Netherlands	1.1	100	136	134	139	143	1.1
Spain	0.9	100	127	149	173	178	1.1
Belgium/Luxembourg	0.9	100	113	103	113	116	0.7
Sweden	0.8	100	113	116	108	114	0.6
Norway	0.5	100	115	124	125	134	0.4
Switzerland	0.4	100	111	122	129	126	0.3
Yugoslavia	0.4	100	135	162	182	192	0.5
Austria	0.4	100	118	125	134	140	0.4
Turkey	0.2	100	126	161	189	189	0.3
Denmark	0.4	100	111	97	104	109	0.3

Cont........

Table 7. continued.......

	% of Total	1969	1972	1975	1978	1979	% of Total
Finland	0.3	100	112	120	138	149	0.3
Greece	0.2	100	148	189	220	227	0.3
Eire	0.1	100	87	90	100	103	0.1
Middle East	1.2	100	133	160	193	179	1.5
N. America	36.3	100	109	108	119	119	30.5
W. Europe	20.8	100	115	115	125	130	19.0
E. Europe	6.3	100	115	125	141	143	6 3
Latin America	3.8	100	118	139	162	169	4.6
S.E. Asia	1.8	100	128	147	189	203	2.6
Africa	1.8	100	111	123	156	166	2.1
S. Asia	1.9	100	106	113	135	139	1.9
Australasia	1.1	100	117	131	150	155	1 2
Eastern Hemisphere	59.8	100	118	130	147	154	64.9
Western Hemisphere	40.1	100	110	111	123	124	35.1
World (Excl. USSR, E. Europe, China)	73.9	100	113	116	130	132	68.9
World	100	100	115	122	138	142	100

Source: BP / Own Calculations

Table 8. THE WORLD'S MAJOR ENERGY CONSUMERS
CONSUMPTION OF PRIMARY FUELS 1979

(Unit: Million tonnes of oil equivalent)

	OIL	GAS	COAL	HYDRO	NUCLEAR	TOTAL
	1979	1979	1979	1979	1979	1979
USA	862.9	498.8	384.1	80.1	72.2	1,898.1
USSR	441.1	307.0	342.5	45.0	12.5	1,148.0
China	91.1	66.8	410.0	9.0	-	576.9
Japan	265.4	22.1	58.6	19.9	14.7	380.7
W. Germany	146.9	45.9	78.6	4.0	9.6	285.0
Canada	89.9	49.2	21.6	54.2	8.5	223.4
UK	94.1	41.2	76.1	1.2	8.1	220.8
France	118.1	23.3	29.9	14.5	9.6	195.4
Italy	101.2	22.9	10.0	12.4	1.3	147.8
Netherlands	38.5	32.9	3.4	-	0.9	75.7
Spain	47.3	1.4	14.5	12.3	1.7	77.2
Belgium/Luxembourg	29.4	10.5	10.2	0.1	2.5	52.7
Sweden	28.4	-	1.8	10.5	3.6	44.3
Norway	9.0	-	0.5	21.8	-	31.3
Switzerland	12.9	0.7	0.5	7.9	2.4	24 4
Yugoslavia	15.8	1.6	13.9	6.6	-	37.9
Austria	12.4	4.3	3.1	6.3	-	26.1
Turkey	15.2	-	5.5	2.4	-	23.1

Contd..........

Table 8.continued ......

	Oil	Gas	Coal	Hydro	Nuclear	Total
	1979	1979	1979	1979	1979	1979
Denmark	16.1	-	4.2	-	-	20.3
Finland	13.3	0.8	3.2	2.7	1.6	21.6
Greece	11.9	-	7.5	1.0	-	20.4
Eire	6.2	-	0.6	0.2	-	7.0
Middle East	74.8	30.0	-	1.0	-	105.8
N. America	952.8	548.0	405.7	134.3	80.7	2,121.5
W. Europe	726.5	185.5	264.0	108.3	41.3	1,325.6
E. Europe	101.1	61.5	270.0	5.4	3.8	441.8
Latin America	211.8	44.0	16.0	46.8	0.8	319.4
S.E. Asia	116.9	7.9	47.8	9.5	1.4	183.5
Africa	63.5	9.0	58.1	12.5	-	143.1
S. Asia	36.7	6.4	74.9	10.7	0.6	129.3
Australasia	38.0	8.4	29.0	9.4	-	84.8
Eastern Hemisphere	1,955.0	704.6	1,534.9	230.7	74.3	4,519.5
World (Exc. USSR, E. Europe, China)	2,486.4	861.3	954.1	352.4	139.5	4,793.7
World	3,119.6	1,296.6	1,976.6	411.8	155.8	6,960.4

Source: BP

Table 9. **THE WORLD'S MAJOR ENERGY CONSUMERS CONSUMPTION OF PRIMARY FUELS 1979**

(Unit Percentage)

	OIL	GAS	COAL	WATER POWER	NUCLEAR ENERGY	TOTAL
	1979	1979	1979	1979	1979	1979
USA	45	26	20	4	4	100
USSR	38	27	30	4	1	100
China	16	12	71	2	-	100
Japan	70	6	15	5	4	100
W. Germany	51	10	28	1	3	100
Canada	40	22	10	24	4	100
UK	43	19	34	1	4	100
France	60	12	15	7	5	100
Italy	68	15	7	8	1	100
Netherlands	51	43	4	-	1	100
Spain	61	2	19	16	2	100
Belgium/Luxembourg	56	20	19	-	5	100
Sweden	64	-	4	24	8	100
Norway	29	-	2	70	-	100
Switzerland	53	3	2	32	10	100

Contd.........

Table 9. continued.........

	OIL	GAS	COAL	WATER POWER	NUCLEAR ENERGY	TOTAL
	1979	1979	1979	1979	1979	1979
Yugoslavia	42	4	37	17	-	100
Austria	47	16	12	24	-	100
Turkey	66	-	24	10	-	100
Denmark	79	-	21	-	-	100
Finland	62	4	15	12	7	100
Greece	58	-	37	5	-	100
Eire	88	-	9	3	-	100
Middle East	71	28	-	1	-	100
N. America	45	26	19	6	4	100
W. Europe	54	14	20	8	3	100
E. Europe	23	14	61	1	1	100
Latin America	66	14	5	15	-	100
S.E.Asia	64	4	26	5	1	100
Africa	44	6	41	9	-	100
S. Asia	28	5	37	8	-	100
Australasia	45	10	34	11	-	100
E. Hemisphere	48	24	17	7	3	100
World (excl. USSR, E. Europe, China)	52	18	20	7	3	100

Source: BP

Table 10. CHANGES IN ENERGY CONSUMPTION 1979–2000

	% change	Average change % P/A		
	1979-2000	1979-2000	1974-1979	1969-1979
USA	+ 55	+1.9	+2.5	+3.9
USSR	+ 72	+2.6	+4.4	+4.9
China	+108	+3.6	+7.9	+8.5
Canada	+ 60	+2.3	+2.5	+3.9
W. Europe	+ 41	+1.6	+1.8	+2.7
Japan	+ 68	+2.5	+1.5	+4.2
Middle East	+358	+7.5	+1.9	+6.0
Africa	+325	+7.1	+7.1	+5.2
Australasia	+ 74	+2.7	+3.9	+4.4
Latin America	+111	+3.6	+4.6	+5.4
E. Europe	+ 60	+2.3	+3.8	+3.6
S. Asia	+120	+3.8	+4.5	+3.4
S.E. Asia	+115	+3.7	+7.3	+7.3
World	+ 77	+2.7	+3.1	+3.6
Communist World*	+ 79	+2.8	+5.1	+5.4
Non-Communist World	+ 75	+2.7	+2.3	+2.8

*Not including Yugoslavia, Cuba.

Source: Own Calculations.

Table 11. WORLD ENERGY CONSUMPTION - THE MAIN CONSUMERS BY PRIMARY FUEL 1979-2000

(Unit: million tonnes oil equivalent)

	1979						2000					
	OIL	COAL	NATURAL GAS	HYDRO	NUCLEAR	TOTAL	OIL	COAL	NATURAL GAS	HYDRO	NUCLEAR	TOTAL
USA	862.9	384.1	498.8	80.1	72.2	1,898.1	916[1]	1,152	481	114	286	2,949
USSR	441.0	342.5	307.0	45.0	12.5	1,148.0	611	712	530	78	47	1,976
China	91.1	410.0	66.8	9.0	-	576.9	122	951	114	13.6	-	1,201
Canada	89.9	21.6	49.2	54.2	8.5	223.4	120[2]	55	78	70	34	357
W. Europe	726.5	264.0	185.5	108.3	41.3	1,325.6	707	457	274	121	308	1,867
Japan	265.4	58.6	22.1	19.9	14.7	380.7	264	90	124	57	105	640
Middle East	74.8	-	30.0	1.0	-	105.8	350	-	134	1.4	1.4	485
Africa	63.5	58.1	9.0	12.5	-	143.1	260	265	56	21	6.4	608
Australasia	38.0	29.0	8.4	9.4	-	84.8	38	60	37	13.0	-	148
Latin America	311.8	16.0	44.0	46.4	0.8	319.4	346	53	102	149.9	23	674
E. Europe	101.1	270.0	61.5	5.4	3.8	441.8	117	463	101	9	17	707
S. Asia	36.7	74.9	6.4	10.7	0.6	129.3	59	180	11	17.3	18	285
S.E. Asia	116.9	47.8	7.9	9.5	1.4	183.5	199	105	39	16.3	36	395
World	3,119.6	1,976.6	1,296.6	411.8	155.8	6,960.4	4,109	4,543	4,543	2,081	681.5	12,296

1 includes synthetics and natural gas liquids.

2 includes synthetics.

Source: Own Calculations.

Table 12.

PRIMARY FUEL CONSUMPTION – MAIN CONSUMERS 1979–2000

(By %)

	1979						2000					
	Oil	Coal	Gas	Hydro & other	Nuclear	Total	Oil	Coal	Gas	Hydro & other	Nuclear	Total
USA	45	20	26	4	4	100	31	39	16	4	10	100
USSR	38	30	27	4	1	100	31	36	27	4	2	100
China	10	71	12	2	-	100	10	79	9	1	-	100
Canada	40	10	22	24	4	100	34	15	22	20	9	100
W. Europe	54	20	14	8	3	100	38	24	15	6	16	100
Japan	70	15	6	5	4	100	41	14	19	9	16	100
Middle East	71	-	28	1	-	100	72	-	28	-	-	100
Africa	44	41	6	9	-	100	43	44	9	3	1	100
Australasia	45	34	10	11	-	100	26	40	25	9	-	100
Latin America	66	5	14	15	-	100	51	8	15	22	3	100
E. Europe	23	61	14	1	1	100	16	65	14	1	2	100
S. Asia	28	57	5	8	-	100	21	63	4	6	6	100
S.E. Asia	64	26	4	5	1	100	50	27	10	4	9	100
World	45	28	19	6	2	100	33	37	17	5	7	100

Source : Own Calculations.

Table 13. WORLD RESOURCES OF FOSSIL FUELS

(billion tonnes of oil equivalent)

	Proven		Probable		Total		Of Which at present usable	
	BT	% of Total	BT	% of Total	BT	% of Total		% of Total
Conventional oil	85	4	241	3	326	3	85	14
Shale oil	33	1	482	6	516	5	20	3
Oil sands	33	1	241	3	275	3	20	3
Natural gas	53	2	185	2	238	2	53	9
Total hydrocarbons	205	9	1,150	15	1,355	13	178	29
Hard coal	1,340	60	5,178	65	6,518	64	330	540
Lignite	670	30	1,607	20	2,277	22	96	16
Total coal	2,010	91	6,785	85	8,795	89	427	71
Grand Total	2,215	100	7,935	100	10,150	100	605	100

Source: H. Rolshoven: General Report summarising World Energy Conference 1980 Papers on energy supply.

Table 14. PROVEN WORLD ENERGY RESERVES 1979

	Reserves (MTOE)	Consumption (MTOE)	Number of Years Until Exhausted
Oil	87,000	3,120	28
Gas	59,000	1,300	45
Coal	390,000	1,980	197
Oil shale/tar sands	45,000	-	-
TOTAL	581,000	6,800	85

Source: Own Calculations. / Various Sources

Table 15. WORLD ENERGY RESERVES – DEPLETION PROSPECTS SELECTED COUNTRIES AND REGIONS 1979

	Oil BTOE	Gas BTOE	Coal[1] BTOE	Total Fuel Reserve (excl. uranium, oil shale, tar sands) BTOE	Total primary consumption (excl. nuclear & hydro) 1979 (MTOE)	Years Potential life[3]
USA	4.2	5.1	75.0	84.3	1,746	49
USSR	9.1	26.2	59.0	91.3	1,090	86
China	2.7	0.6	56.3	59.6	568	105
E. Europe	0.4	0.3	18.5	19.5	433	45
Australia	-	0.7	16.1	16.8	75[2]	215
S. Africa	-	-	13.4	13.4	-	-
India	-	-	8.0	8.0	-	-
W. Germany	-	0.2	5.4	5.6	271	20
S. America	7.9	3.5	2.7	14.1	272	52
UK	2.1	0.7	2.7	5.5	211	26
Canada	0.9	2.3	-	3.2	161	20
W. Europe	3.2	3.4	8.1	14.7	1,176	12
Iran	7.9	12.0	-	19.9	-	-
Africa	7.6	4.4	16.1	28.1	131	214
S. Arabia	22.2	1.8	-		-	-
Middle East	49.2	18.1	-	67.3	105	641
Non-Communist World	75.1	31.9	133.9	362.6	4,302	57
Communist World	12.2	27.1	134.1	173.4	2,091	80
TOTAL	87.3	59	390	536	6,393	65

1 economically recoverable hard coal. 2 Australasia. 3 At 1979 consumption rate

Source: Own Calculations. / WOCOL / BP / Le Petrole en Chiffres / Cedigaz / Others

Table 16. WORLD ECONOMICALLY RECOVERABLE COAL RESERVES (HARD COAL)

(Unit: million tonnes of oil equivalent)	MTOE	% of Total
USA	75,000	28
USSR	58,960	22
China	56,280	21
Eastern Europe (mainly Poland)	18,760	7
Australia	16,080	6
South Africa	13,400	5
India	8,040	3
West Germany	5,360	2
South America	2,680	1
U.K.	2,680	1
Others	10,720	4
TOTAL	268,000	100

Notes: BTOE

Estimated world coal resources = 6,500

Economically recoverable at present = 390

Recoverable hard coal (bituminous coal and anthracite) = 268.

Source: BP.

Table 17. **WORLD COAL – GEOLOGICAL RESOURCES BY COUNTRY**

Country	Geological Resources (MTOE)	% of Total
USSR	3,256,200	45.0
USA	1,722,167	24.0
China	963,490	13.0
Australia	402,000	5.6
Canada	216,434	3.0
West Germany	165,356	2.3
U.K.	127,300	1.8
South Africa	48,240	0.7
India	54,283	0.7
Others	153,540	2.1
TOTAL	7,202,642	100.0

Source: World Energy Conference.

Table 18. WORLD COAL CONSUMPTION BY REGION 1969–79

(Unit: million tonnes of oil equivalent)

(MTOE)	% of total	1969	1970	1971	1972	1973	1974	1975	1976	1977	1978	1979	% of total
Latin America	0.6	9.5	10.2	10.5	11.3	11.7	13.1	14.2	15.2	14.1	15.2	16.0	0.8
North America	21.2	342.9	346.4	332.4	331.8	350.6	347.8	338.4	364.8	379.7	374.4	405.7	20.5
Western Europe	18.7	302.7	292.4	269.1	241.9	254.8	269.2	236.2	249.3	248.2	248.7	264.0	13.4
Middle East		†	-	-	-	-	-	-	-	-	-	-	-
Africa	2.5	40.2	41.4	44.4	42.5	38.6	40.4	43.1	46.7	47.3	53.3	58.1	2.9
South Asia	3.4	54.5	51.7	52.4	54.6	55.0	56.8	59.1	62.5	64.0	67.2	74.9	3.8
South East Asia	1.8	29.3	31.9	33.8	35.5	36.4	38.0	40.4	39.8	43.1	47.2	47.8	2.4
Eastern Europe	14.1	227.4	236.9	241.2	241.3	235.0	237.6	246.9	253.9	260.7	266.7	270.0	13.7
Australasia	1.3	21.2	21.5	21.3	22.1	22.9	24.4	26.1	24.9	26.8	28.2	29.0	1.5
Western Hemisphere	21.8	352.4	456.6	342.9	343.1	362.3	360.9	352.6	380.0	393.8	389.6	421.7	21.3
Eastern Hemisphere	78.2	1,262.2	1,284.3	1,293.8	1,295.1	1,311.1	1,333.8	1,361.9	1,410.2	1,443.0	1,489.7	1,554.9	78.7
World (excl. USSR, E. Europe & China)	53.3	860.8	858.1	820.0	801.8	830.9	833.5	819.6	863.1	880.5	888.2	954.1	48.3
Communist World	46.7	753.8	782.8	816.7	802.4	842.5	861.2	894.9	827.1	956.3	991.1	1,022.5	51.7
World	100	1,614.6	1,640.9	1,636.7	1,638.2	1,673.4	1,694.7	1,714.5	1,790.2	1,836.8	1,879.3	1,976.6	100

Source: BP.

Table 19. WORLD COAL CONSUMPTION BY MAIN CONSUMING COUNTRIES 1969–79

(Unit: million tonnes of oil equivalent)

	1969	1970	1971	1972	1973	1974	1975	1976	1977	1978	1979	% of total %	% Change 1969–79 %
China	277.4	252.0	273.2	278.1	292.5	307.4	321.8	337.5	354.4	380.0	410.0	21	+ 80
USA	327.1	329.5	316.3	316.6	335.0	331.9	322.9	346.5	356.3	355.2	384.1	19	+ 17
USSR	299.0	293.9	302.3	317.0	315.0	316.2	326.2	335.7	341.2	344.4	342.5	17	+ 14
W. Germany	92.2	89.6	83.8	80.2	82.1	82.5	70.7	75.9	71.7	72.9	78.6	4	– 15
UK	96.3	92.3	82.9	72.2	78.4	69.1	71.9	72.7	73.4	70.4	76.1	4	– 21
Japan	60.5	62.6	56.1	57.1	60.9	63.8	62.1	39.9	57.3	54.0	58.6	3	– 3
France	39.8	37.8	33.8	29.5	29.5	30.3	26.5	30.0	29.8	30.5	29.9	1	– 25
Canada	15.8	16.9	16.1	15.2	15.6	15.9	15.5	18.3	23.4	19.2	21.6	1	+ 37
Spain	10.3	10.8	10.9	11.1	10.8	11.3	11.7	12.2	13.4	13.9	14.5	1	+ 41
Yugoslavia	9.1	9.7	10.7	10.4	11.4	11.7	12.6	12.6	12.5	12.8	13.9	1	+ 53
Belgium/Luxembourg	17.3	16.4	13.7	13.1	11.0	11.9	8.9	9.3	9.4	9.7	10.2	-	– 41
Italy	9.8	9.9	9.5	8.9	9.0	9.8	9.8	9.7	9.6	9.8	10.0	-	+ 2
Greece	2.2	2.5	3.7	4.0	4.7	4.9	6.5	7.6	8.1	7.2	7.5	-	+241
Turkey	4.2	4.2	4.3	4.5	4.6	4.8	4.9	5.3	5.5	5.4	5.5	-	+ 31
Denmark	2.2	2.0	1.4	1.3	1.8	2.1	2.5	2.5	3.3	3.4	4.2	-	+ 91
Netherlands	6.7	4.8	3.6	3.1	3.1	2.9	2.5	3.2	3.2	3.1	3.4	-	– 49
Finland	2.6	2.5	2.2	1.8	2.0	2.1	1.8	2.5	2.6	3.5	3.2	-	+ 23
Austria	3.8	4.0	3.5	3.6	3.6	3.4	3.2	3.1	2.9	2.8	3.1	-	– 18
Sweden	1.1	1.1	0.9	0.8	0.8	0.6	0.9	1.2	1.0	1.5	1.8	-	+ 64
Eire	2.9	2.7	2.5	0.8	0.7	0.7	0.7	0.4	0.6	0.6	0.6	-	– 79
Switzerland	0.6	0.6	0.4	0.4	0.2	0.1	0.1	0.2	0.3	0.3	0.5	-	– 17
Norway	1.1	1.0	0.9	0.8	0.5	0.6	0.6	0.5	0.5	0.5	0.5	-	– 54
Portugal	0.5	0.5	0.4	0.4	0.6	0.4	0.4	0.4	0.4	0.4	0.5	-	-
TOTAL	1,614.6	1,640.9	1,636.7	1,638.2	1,673.4	1,694.7	1,714.5	1,790.2	1,836.8	1,879.3	1,976.6	100	+ 22

Source: BP.

Table 20. **COAL CONSUMPTION 1979-2000 - WESTERN EUROPE**

	1979		1990		2000		
	MTOE	% of Total	MTOE	% of Total	MTOE	% of Total	% Change 1979-2000
Denmark	4.2	1.6	6	1.6	7	1.6	+ 66
Finland	3.2	1.2	5	1.4	7	11.6	+ 119
France	29.9	11.3	37	9.9	40	8.6	+ 34
West Germany	78.6	29.7	105	28.3	116	26.9	+ 48
Italy	10.0	3.8	17	4.7	27	5.7	+ 170
Netherlands	3.4	1.3	10	2.6	19	4.0	+ 459
Sweden	1.8	0.7	7	1.85	14	3.0	+ 678
U.K.	76.1	28.8	98	26.4	112	24.0	+ 47
Others	56.8	21.5	86	23.0	114	24.5	+ 101
TOTAL	264.0	100.0	372	100.0	457	100.0	+ 73

Source: Own Calculations

Table 21. **WORLD COAL CONSUMPTION 1979–2000**

	1979		1990		2000		% change
	MTOE	% of Total	MTOE	% of Total	MTOE	% of Total	1979–2000
USA	384.1	19.4	690	22.3	1,152	25.4	+200
USSR	342.5	17.3	486	15.7	712	15.7	+108
China	410.0	20.7	689	22.2	951	20.9	+132
Canada	21.6	1.1	30	1.2	55	1.2	+155
W. Europe	264.0	13.4	372	12.0	457	10.1	+ 73
Japan	58.6	3.0	69	2.2	90	2.0	+53.6
Africa	58.1	2.9	129	4.2	265	5.8	+350
Australasia	29.0	1.5	42	1.3	60	1.3	+107
Latin America	16.0	0.8	36	1.2	53	1.2	+231
E. Europe	270.0	13.7	358	11.6	463	10.2	+ 71
S. Asia	74.9	3.8	118	3.8	180	4.0	+140
S.E. Asia	47.6	2.4	66	2.1	105	2.3	+120
Non-Communist World	954.1	48.3	1,558	50.4	2,417	53.2	+153
Communist World	1,022.5	51.7	1,533	69.1	2,126	116.8	+108
World	1,976.6	100	3,091	100	4,543	100	+130

Source: Own Calculations / BP

Table 22. COAL USE – 1977 AND 2000 – OECD COUNTRIES (%)

	1977				2000			
	Industry Markets	Electric Markets	Other (domestic etc.)	Total	Industry Markets	Electric Markets	Other (domestic (etc.)	Total
Canada	4	67	29	100	13	60	27	100
USA	12	73	15	100	12	74	14	100
Denmark	17	83	-	100	8	92	-	100
Finland	17	43	40	100	17	33	50	100
France	7	49	44	100	11	22	69	100
W. Germany	4	59	37	100	5	66	29	100
Italy	1	13	86	100	9	52	39	100
Netherlands	7	23	70	100	13	60	27	100
Sweden	20	-	80	100	18	36	46	100
UK	8	60	32	100	21	53	26	100
Other W. Europe	9	47	44	100	5	63	32	100
Japan	6	7	87	100	5	38	57	100
Australia	12	67	21	100	12	77	11	100
TOTAL OECD	9	61	30	100	11	66	23	100

Source: Own Estimates / WOCOL / Others

Table 23. COAL FIRED PROPORTION OF ELECTRICAL CAPACITY

	1977 %	2000 %
Japan	4	16
Western Europe	36	36
North America	36	48
Australia	64	84
TOTAL OECD	32	39

Source: Own Estimates / WOCOL / Others

Table 24. **COAL PRODUCTION 1977 and 2000**

(million tonnes of oil equivalent)

	1977		2000	
	MTOE	% of Total	MTOE	% of Total
USA	375	21.5	1.262	27.8
China	354	20.2	971	21.4
USSR	342	19.6	737	16.2
Australia	51	2.9	218	4.8
Poland	112	6.4	208	4.6
India	48	2.7	190	4.2
South Africa	49	2.8	153	3.4
U.K.	72	4.1	93	2.0
Canada	15	0.9	106	2.3
West Germany	80	4.6	100	2.2
Indonesia	1	-	13	0.3
Japan	12	0.7	12	0.3
France	14	0.8	7	0.1
Italy	1	-	2	0.1
Denmark	-	-	1	0.1
OECD Europe	192	11.0	285	6.3
Africa/Latin America	17	1.0	136	3.0
Other centrally planned economies	167	9.6	251	5.5
North America	380	21.8	1.368	30.1
Other Asia	10	0.6	7	0.1
WORLD	1.745	100.0	4.543	100.0

Source: World Energy Conference / Own Estimates

Table 25. **TOTAL COAL IMPORTS 1977 AND 2000**

(million tonnes oil equivalent)

	1977		2000		
	MTOE	% of Total	MTOE	% of Total	% Change
Denmark	3	2.4	6	1.4	+ 100
Finland	3	2.4	6	1.4	+ 100
France	16	12.6	33	8.0	+ 106
West Germany	3	2.4	32	7.7	+ 967
Italy	9	7.1	25	6.0	+ 178
Netherlands	3	2.4	17	4.1	+ 467
Sweden	1.4	1.1	12	2.9	+ 757
U.K.	1	0.8	4	1.0	+ 300
Other Western Europe	9	7.1	37	9.0	+ 311
OECD Europe	48	37.8	172	41.6	+ 258
Canada	9	7.1	11	2.7	+ 22
Japan	41	32.3	88	21.3	+ 115
Total OECD	97	76.4	271	65.6	+ 179
East and other Asia	2	1.6	67	16.2	+ 3,250
Africa and Latin America	5	3.9	42	10.2	+ 740
Centrally planned economies	23	18.1	33	8.0	+ 43
TOTAL WORLD	127	100.0	413	100.0	+ 225

Source: WOCOL / Own Estimates

Table 26. METALLURGICAL COAL IMPORTS BY COUNTRY AND REGION 1977 AND 2000

(million tonnes of oil equivalent)	1977		2000		
	MTOE	% of Total	MTOE	% of Total	% Change
Finland	0.6	0.7	0.7	0.4	+ 17
France	7	8.0	12	6.5	+ 71
West Germany	0.7	0.8	-	-	-
Italy	7	8.0	11	6.0	+ 57
Netherlands	2	2.3	3	1.6	+ 50
Sweden	1.2	1.4	2	1.1	+ 67
U.K.	0.7	0.8	4	2.2	+ 471
Other Western Europe	4	4.6	16	8.7	+ 300
OECD Europe	23	26.4	48	26.1	+ 109
Canada	5	5.7	6	3.3	+ 20
Japan	40	46.0	53	28.8	+ 32
Total OECD	67	77.0	97	52.7	+ 45
East and other Asia	2	2.3	27	14.7	+ 1,250
Africa and Latin America	5	5.7	38	20.6	+ 660
Centrally planned economies	11	12.6	13	7.1	+ 18
TOTAL WORLD	87	100.0	184	100.0	+ 111

Source: WOCOL / Own Estimates

Table 27. **WORLD STEAM COAL IMPORTS 1977 AND 2000 SELECTED COUNTRIES AND REGIONS**

(million tonnes of oil equivalent)	1977		2000		
	MTOE	% of Total	MTOE	% of Total	% Change
Japan	1.3	3.2	35.5	17.7	+ 34.2
France	9.4	23.3	21.4	8.7	+ 12.0
West Germany	2.0	5.0	32.4	6.7	+ 30.4
Italy	1.3	3.2	14.0	5.5	+ 12.7
Netherlands	1.0	2.5	14.3	6.6	+ 13.3
Sweden	0.2	0.5	9.6	4.8	+ 9.4
Denmark	3.1	7.7	6.3	3.1	+ 3.2
Canada	4.0	9.9	5.4	2.7	+ 1.4
Finland	2.7	6.7	5.2	2.6	+ 2.5
Other Western Europe	5.0	12.4	21.4	10.6	+ 16.4
OECD Europe	24.8	61.7	124.6	48.7	+ 99.8
Total OECD	30.1	74.9	165.5	70.0	+ 135.4
East and other Asia	-	-	40.2	20.0	+ 40.2
Africa and Latin America	0.7	1.7	4.0	2.0	+ 3.3
Centrally planned economies	11.4	28.3	20.1	10.0	+ 8.7
TOTAL WORLD	40.2	100.0	229.8	100.0	+ 189.6

Source: WOCOL / Own Estimates

Table 28. **TOTAL COAL EXPORTS 1977 AND 2000**

(million tonnes of oil equivalent)	1977		2000		
	MTOE	% of Total	MTOE	% of Total	% Change
USA	32.1	25.3	110	26.6	+ 243
Australia	24.8	19.5	103	24.9	+ 315
South Africa	7.3	5.7	39	9.4	+ 434
Canada	7.3	5.7	62	15.0	+ 749
Poland	25.4	20.0	27	6.5	+ 6
USSR	16.0	12.6	27	6.5	+ 69
China	1.3	1.0	14	3.4	+ 977
West Germany	8.7	6.8	16	3.9	+ 84
India and Indonesia	0.7	0.5	3	0.7	+ 329
Latin America and other Africa	4.0	3.1	12	2.9	+ 200
WORLD TOTAL	127.0	100.0	413	100.0	+ 225

Source: WOCOL / Own Estimates

Table 29. THE WORLD COAL MARKET IN THE YEAR 2000

(million tonnes oil equivalent)

	Production	Imports	Exports	Consumption
USA	1,262	-	110	1,152
Canada	106	11	62	55
Australia	218	-	103	60*
Italy	2	25	-	27
Netherlands	-	17	-	17
U.K.	93	4	-	97
France	7	42	-	49
Germany	100	32	16	116
Western Europe	285	172	-	457
USSR	737		27	712
Poland	208	] 33	27	] 463
Eastern Europe	251		-	
China	971		14	951
South Asia	] 190		-	] 180
India		] 67	] 3	
Indonesia	13			] 105
SE Asia	-		-	
South Africa	153		39	265
Other Africa	] 136	] 42	12	
Latin America				53
Japan	12	88	-	90
WORLD	4,543	413	413	4,543

Source: Own Estimates / Various Sources

* Australasia.

Table 30. **GAS CONSUMPTION 1969–1979 (by Region)**

(Unit: Million tonnes oil equivalent)

	N. America MTOE	Latin America MTOE	W. Hemis- MTOE	W. Europe MTOE	E. Europe MTOE	E. Hemis- MTOE	TOTAL		
							MTOE	% of Total Energy	Index (1969 =100)
1969	590.8	34.1	624.9	53.8	30.1	252.6	877.5	18	100
1970	596.8	30.4	627.2	72.6	33.3	301.5	928.7	18	106
1971	619.1	32.6	651.7	93.4	38.5	345.9	997.6	18	114
1972	626.7	36.4	663.1	113.5	41.2	385.3	1,048.4	18	119
1973	614.1	36.5	650.6	129.9	42.1	431.4	1,082.0	18	123
1974	597.3	37.8	653.1	147.3	46.0	476.6	1,111.7	19	127
1975	551.8	39.2	591.0	153.5	51.3	513.6	1,104.6	19	126
1976	562.5	40.5	603.0	163.6	57.5	563.0	1,166.0	19	133
1977	548.2	39.6	587.8	169.9	58.4	599.6	1,187.4	19	135
1978	551.5	42.3	593.8	174.6	60.0	637.6	1,231.4	19	140
1979	548.0	44.0	592.0	185.5	61.5	704.6	1,296.6	19	148
Yearly change 1979 over 1969	−0.8	+2.6	−0.5	+13.2	+7.4	+10.8	+3.6		
Yearly change 1970 over 1974	−1.7	+3.1	−1.4	+4.7	+6.0	+8.1	+3.1		

BP | Own Calculations

Table 31. NATURAL GAS MAIN CONSUMERS - ENERGY BALANCE 1979

(million tonnes oil equivalent)

	Production MTOE	Consumption MTOE	Balance MTOE
USA	480.3	498.8	– 18.5
USSR	350.0	307.0	+ 43.0
China	68.9	66.8	+ 2.1
Canada	63.2	49.2	+ 14.0
West Germany	17.8	45.9	– 28.1
U.K.	34.2	41.2	– 7.0
Netherlands	80.3	32.9	+ 47.4
Middle East	31.7	30.0	+ 1.7
France	6.7	23.3	– 16.6
Italy	11.6	22.9	– 11.3
Japan	2.3	22.1	– 19.8
Belgium/Luxembourg	-	10.5	– 10.5
Africa	25.5	9.0	+ 16.5
Australia	8.0	8.4	– 0.4
WORLD	1,337.1	1,296.6	–

Source: Own Calculations / Various Sources

Table 32. WORLD NATURAL GAS CONSUMPTION BY MAIN CONSUMING COUNTRIES 1969-79
(Unit: million tonnes of oil equivalent)

	1969	1970	1971	1972	1973	1974	1975	1976	1977	1978	1979	% of total	Yearly % Change 1974–1979
USA	561.2	564.1	584.2	587.4	572.3	555.1	508.7	516.4	302.3	504.2	498.8	38.4	– 2.1
Canada	29.6	32.7	34.9	39.3	41.8	42.2	43.1	46.1	45.9	47.3	49.2	3.8	+ 3.1
Austria	2.2	2.5	2.9	3.1	3.4	3.7	3.6	4.1	4.2	4.4	4.3	0.3	+ 3.1
Belgium/Luxembourg	2.6	4.2	5.8	6.7	8.2	9.8	9.6	10.3	10.1	9.9	10.5	0.8	+ 1.4
Finland	-	-	-	-	-	0.4	0.7	0.8	0.7	0.8	0.8	-	+16.6
France	8.5	9.3	11.1	13.2	15.7	17.2	17.0	19.0	20.4	20.9	23.3	1.8	+ 6.3
Italy	11.2	12.3	12.5	12.3	14.4	15.8	18.0	22.0	21.6	22.5	22.9	1.8	+ 7.7
Netherlands	13.4	19.1	24.6	29.0	32.2	32.1	33.2	33.0	33.4	32.6	32.9	2.5	+ 0.5
Spain	0.1	0.1	0.4	1.0	1.0	1.3	1.3	1.5	1.4	1.5	1.4	0.1	+ 2.4
Switzerland	-	-	-	0.1	0.2	0.4	0.6	0.5	0.6	0.7	0.7	-	+15.9
UK	5.9	11.2	18.1	25.2	26.1	31.8	32.9	34.6	36.9	37.9	41.2	3.2	+ 5.3
W. Germany	9.2	13.0	16.9	21.5	27.0	32.5	34.4	36.3	38.9	41.7	45.9	3.5	+ 7.1
Yugoslavia	0.7	0.9	1.1	1.4	1.7	2.3	2.2	1.5	1.7	1.7	1.6	0.1	– 6.9
Middle East	11.3	19.0	19.0	21.0	24.1	27.6	26.2	26.9	28.8	30.1	30.0	2.3	+ 1.7
Africa	1.4	1.5	1.7	2.4	3.2	3.6	4.3	5.1	6.9	8.3	9.0	0.7	+20.0
S. Asia	4.8	5.4	6.0	6.8	8.0	7.9	8.1	8.8	9.4	9.7	6.4	0.5	– 4.1
S. East Asia	2.0	2.3	2.3	2.4	3.6	4.3	4.1	4.2	4.6	7.9	7.9	0.6	+12.9
Japan	2.4	3.6	3.8	3.7	5.3	7.0	8.5	10.4	13.0	17.0	22.1	1.7	+25.7
Australasia	0.2	1.5	2.2	3.2	3.9	4.6	4.9	6.1	7.4	7.8	8.4	0.6	+12.8
USSR	144.7	159.0	173.9	182.8	198.8	210.8	230.0	253.1	271.2	289.2	307.0	25.6	+ 7.8
China	1.9	3.3	5.2	8.3	12.5	17.5	22.7	27.3	30.0	33.0	66.8	5.1	+30.8
E. Europe	30.1	33.3	38.5	41.2	42.1	46.0	51.3	57.5	58.4	60.0	61.5	4.7	+ 6.0
World (excl. USSR, China & E. Europe)	700.8	733.1	780.1	816.1	828.6	837.4	800.6	828.1	827.8	849.2	861.3	66.4	+ 0.6
World	877.5	928.7	997.6	1,048.4	1,082.0	1,111.7	1,104.6	1,166.0	1,187.4	1,231.4	1,296.6	100	+ 3.1

Source: BP

Table 33. **NATURAL GAS CONSUMPTION 1979–2000**

(million tonnes oil equivalent)

	1979		1990		2000		
	MTOE	% of Total	MTOE	% of Total	MTOE	% of Total	% Change 1979-2000
USA	498.8	38.5	496	27.6	481	23.1	– 3.6
USSR	307.0	23.7	505	28.1	530	25.5	+ 72.6
China	66.8	5.1	113	6.3	114	5.5	+ 70.7
Canada	49.2	3.8	66	3.7	78	3.7	+ 58.5
Western Europe	185.5	14.3	257	14.3	274	13.2	+ 47.7
Japan	22.1	1.7	52	2.9	124	6.0	+461.1
Middle East	30.0	2.3	79	4.4	134	6.4	+346.7
Africa	9.0	0.7	35	1.9	56	2.7	+522.2
Australasia	8.4	0.6	19	1.1	37	1.8	+340.0
Latin America	44.0	3.4	67	3.7	102	4.9	+131.8
Eastern Europe	61.5	4.7	84	4.7	101	4.8	+ 64.2
South Asia	6.4	0.5	7	0.4	11	0.5	+ 71.9
SE Asia	7.9	0.6	19	1.1	39	1.9	+393.7
World excl. USSR, China & E. Europe	861.3	66.4	1,097	61.1	1,330	64.2	+ 55.1
Communist World	435.3	33.6	702	39.0	745	35.8	+ 71.1
WORLD	1,296.6	100.0	1,799	100.0	2,081	100.0	+ 60

Source: Own Calculations.

Table 34. **NATURAL GAS – PROVEN RESERVES 1970/1980**

	MTOE (Jan 1970)	% of Total	MTOE (Jan 1980)	% of Total	Life of Reserves at Present Production Rate (Years)
USSR	**7,842**	**23.9**	**26,316**	**41.4**	**75**
Iran	2,606	7.9	9,202	14.5	973
USA	6,700	20.4	4,747	7.5	10
Algeria	3,531	10.8	2,322	3.7	126
Canada	1,267	3.9	2,147	3.4	34
Saudi Arabia	1,288	3.9	1,803	2.8	349
Mexico	292	0 9	1,490	2.3	125
Netherlands	2,082	6.3	1,449	2.3	18
Nigeria	122	0.4	1,251	2.0	728
Venezuela	646	2.0	1,023	1.6	66
Kuwait	1,045	3.2	970	1.5	189
Indonesia	68	0.2	946	1.5	39
Qatar	178	0.5	755	1.2	549
Australia	307	0.9	728	1.1	109
Iraq	475	1.4	660	1.0	549
China	88	0.3	645	1.0	52
Norway	-	-	614	1.0	34
U.K.	852	2.6	609	1.0	18
Libya	633	1.9	598	0.9	154
Abu Dhabi	214	0.6	550	0.9	120
Argentina	156	0.5	396	0.6	53
Pakistan	460	1.4	384	0.6	72
WORLD	32,765	100.0	63,509	100.0	50
Of which OPEC	1 ,852	33.1	20,199	31.8	221

Source: Cedigaz / UN Monthly Bulletin of Statistics / Own Estimates

Table 35. NATURAL GAS – ESTIMATED RESERVES 1980

(million tonnes oil equivalent)

	Proved Reserves MTOE	% of Total	Undiscovered MTOE	TOTAL	
				MTOE	% of Total
USA	4,746	7.5	2,238	6,984	4.2
Canada	2,147	3.4	537	2,680	6.4
Other W. Hemisphere	3,476	5.5	2,242	5,718	3.3
Western Europe	3,218	5.1	5,062	8,280	5.0
Iran	9,202	14.5	12,398	21,600	12.9
Other Middle East	5,256	8.3	9,144	14,400	8.6
Africa (inc. Algeria)	4,418	7.0	8,364	12,782	5.0
Asia Pacific	2,959	4.7	7,841	10,800	6.5
USSR	26,316	41.4	52,204	78,520	44.0
China	645	1.0	6,555	7,200	4.3
TOTAL	63,509	100.0	46,309	172,000	100.0

Source: Cedigaz / Various Sources

(In addition some 24 billion tonnes of natural gas liquids are considered to exist of which 35% are classified proven or probable).

Table 36. NATURAL GAS – MAIN PRODUCERS

(million tonnes oil equivalent)

	Production				% Change
	1978	% of Total	1979	% of Total	1978–1979
USA	486.4	39.5	480.3	37.5	– 1.2
USSR	320.2	26.0	350.0	27.3	+ 9.3
Netherlands	75.7	6.1	80.3	6.3	+ 6.1
Canada	58.5	4.8	63.2	4.9	+ 8.0
China	34.8	1.0	68.9	1.0	+ 5.9
U.K.	33.7	2.7	34.2	2.7	+ 1.5
Indonesia	20.0	1.6	24.2	1.9	+21.0
Romania	24.9	2.0	24.1	1.9	– 3.2
Algeria	12.1	1.0	18.4	1.4	+52.0
Norway	11.7	0.9	18.0	1.4	+53.8
West Germany	17.5	1.4	17.8	1.4	+ 1.7
Venezuela	12.9	1.0	15.5	1.2	+20.1
Mexico	9.2	0.7	12.0	0.9	+30.4
Italy	11.8	1.0	11.6	0.9	– 1.7
Iran	16.1	1.3	9.46	0.7	– 4.2
Brunei	7.6	0.6	7.5	0.6	– 1.3
Argentina	6.8	0.5	7.5	0.6	+10.3
East Germany	7.3	0.6	7.3	0.6	-
Australia	5.9	0.5	6.7	0.5	+13.6
France	6.8	0.5	6.7	0.5	– 1.5
Poland	6.9	0.6	6.3	0.5	– 8.7
Hungary	6.3	0.5	5.6	0.4	–11.1
Saudi Arabia*	4.3	0.3	5.2	0.4	+20.9
Kuwait*	5.4	0.4	5.2	0.4	– 3.7
Abu Dhabi	4.5	0.4	4.6	0.4	+ 7.2
WORLD TOTAL	1,253.0	100.0	1,337.1	100.0	+ 4.1
Of which OPEC	83.6	6.8	91.6	100.0	+ 9.6

Source: Cedigaz / UN Monthly Bulletin of Statistics / Own Calculations.
(*Kuwait, Saudi Arabia include half share of partitioned zone).

Table 37. MIDDLE EAST – NATURAL GAS PRODUCTION AND RESERVES 1978–1979

(million tonnes oil equivalent)

	Production		**Reserves**		
	1978 MTOE	**1979 MTOE**	**1970 MTOE**	**1980 MTOE**	**Life of Reserves (Years)**
Abu Dhabi	3.60	4.50	224	250	122
Bahrain	3.20	3.40	9	243	71
Dubai	0.60	3.40	17	39	68
Iran	16.10	9.50	260	9,202	973
Iraq	1.50	1.30	475	660	549
Israel	0.04	0.04	2	1	17
Kuwait	5.40	5.20	1,045	970	188
Oman	0.34	0.43	49	120	280
Qatar	1.38	1.40	158	755	549
Saudi Arabia*	4.30	5.20	1,288	1,802	349
Syria	0.17	1.50	12	116	670
Turkey	0.09	0.09	8	1	10
TOTAL	36.72	33.06	5,913	14,458	438

Source: Cedigaz / UN Bulletin of Statistics / Own Calculations.

*Kuwait and Saudi Arabia include half shares of neutral zone.

Table 38. FAR EAST – NATURAL GAS PRODUCTION AND RESERVES 1978–1979

(million tonnes oil equivalent)

	Production		Reserves		
	1978 MTOE	**1979 MTOE**	**1970 MTOE**	**1980 MTOE**	**Life of Reserves (Years)**
Afghanistan	2.30	2.10	124	60	29
Australia	5.90	6.60	307	728	110
Bangladesh	0.90	1.05	n.a.	228	217
Brunei	7.60	7.50	49	181	24
Burma	0.30	0.34	3	3	10
China	n.a.	n.a.	88	645	n.a.
India	1.50	1.60	34	224	140
Indonesia	20.00	24.20	68	946	39
Japan	2.50	2.30	17	15	7
Malaysia	0.09	0.09	49	413	15,700
New Zealand	1.10	1.32	158	163	123
Pakistan	5.00	5.30	460	384	72
Taiwan	2.00	2.00	24	22	11
Thailand	-	-	n.a.	185	n.a.
TOTAL	61.00	67.00	1,381	4,235	n.a.

Source: Cedigaz / UN Monthly Bulletin of Statistics.

Table 39. **NORTH AND SOUTH AMERICA – NATURAL GAS PRODUCTION AND RESERVES 1978–1979**

(million tonnes oil equivalent)

	Production		**Reserves**		
	1978 MTOE	**1979 MTOE**	**1970 MTOE**	**1980 MTOE**	**Life of Reserves (Years)**
North America:					
Canada	58.50	63.20	1,266	2,147	34
USA	486.40	480.30	6,699	4,747	10
TOTAL	544.90	543.50	7,965	6,893	13
South America:					
Argentina	6.80	7.50	156	396	53
Barbados	0.01	0.01	-	-	-
Bolivia	1.38	1.40	73	131	94
Brazil	0.97	0.95	22	39	41
Colombia	3.60	3.60	68	103	28
Chile	3.00	2.60	68	60	24
Cuba	0.02	0.02	-	-	-
Ecuador	0.03	0.03	17	37	108
Mexico	9.20	12.04	292	1,490	125
Peru	0.77	0.99	4	26	27
Trinidad	2.15	1.77	73	172	100
Venezuela	12.90	15.50	65	1,023	66
TOTAL	41.10	46.40	1,419	3,746	372

Source: Cedigaz / UN Monthly Bulletin of Statistics.

Table 40. AFRICA – NATURAL GAS PRODUCTION AND RESERVES 1978–1979

(million tonnes oil equivalent)

	Production		Reserves		
	1978 MTOE	1979 MTOE	1970 MTOE	1980 MTOE	Life of Reserves (Years)
Algeria	12.10	18.40	3,530	2,322	126
Angola	0.09	0.09	23.6	34	364
Congo	0.01	0.01	-	26	-
Egypt	0.63	0.95	34	73	77
Gabon	0.05	0.05	13	22	400
Libya	4.30	3.90	632	598	154
Morocco	0.06	0.07	1	1	13
Nigeria	0.46	1.72	123	1,251	728
Rwanda	0.01	0.01	-	-	-
Sudan	-	-	-	3	-
Tanzania	-	-	-	1	-
Tunisia	0.25	0.28	12	86	303
Zaire	-	-	-	3	-
TOTAL	18.00	25.50	4,368	4,418	173

Source: Cedigaz / UN Monthly Bulletin of Statistics.

Table 41. USSR & EASTERN EUROPE – NATURAL GAS PRODUCTION AND RESERVES 1970/80

(Unit: Million tonnes oil equivalent)

	PRODUCTION		RESERVES		
	1978	1979	MTOE		Life of Reserves (Years)
	MTOE	MTOE	1970	1980	
Albania	0.26	0.26	8	10	38
Bulgaria	0.02	0.02	24	4	215
Czechoslovakia	0.82	0.76	12	11	15
E. Germany	7.30	7.30	12	69	9
Hungary	6.30	5.60	102	99	18
Poland	6.90	6.30	8	108	17
Rumania	24.90	24.10	146	116	5
USSR	320.00	350.00	7,842	26,316	75
Yugoslavia	1.70	1.60	29	52	32
TOTAL	368.00	396.00	8,184	26,785	

Source: Cedigaz / UN Monthly Bulletin of Statistics

Table 42. WESTERN EUROPE – NATURAL GAS PRODUCTION AND RESERVES 1978–1979

(million tonnes oil equivalent)

	Production		Reserves	
	1978	**1979**	**1970**	**1980**
Austria	1.80	1.80	9	10
Belgium	0.02	0.02	-	-
Denmark	-	-	-	99
France	6.70	6.70	176	80
West Germany	17.40	17.80	251	157
Greece	-	-	-	9
Ireland	0.09	0.12	-	26
Italy	11.80	11.60	170	155
Netherlands	75.70	80.20	2,081	1,449
Norway	11.60	18.10	-	615
Spain	0.01	0.01	-	9
U.K.	33.70	34.20	851	608
TOTAL	159.10	170.50	3,541	3,218

Source: Cedigaz / UN Monthly Bulletin of Statistics.

Table 43. **TRADE IN LNG 1980**

(million tonnes oil equivalent)

Exporting Country	Market	Year Begun	Contract Volume[2] (MTOE)
Algeria	U.K.	1964	0.88
Algeria	France (Le Havre)	1965	0.44[1]
USA, Alaska	Japan	1969	1.23
Libya	Italy	1969	2.07
Libya	Spain	1969	9.68
Algeria	France (Fos)	1972	3.08[1]
Brunei	Japan	1972	6.60
Algeria	Spain	1976	3.96 (0.88 in 1980)
Abu Dhabi	Japan	1977	3.20
Indonesia	Japan	1977	9.30 (10.38 in 1980)
Algeria	USA (Boston	1978	1.06[1]
Algeria	USA (Cove Point)	1978	8.80[1]
TOTAL			41.58
Suspended shipments			13.38
Actual total			28.20

Source: Paper by E.K. Faridany, Ocean Phoenix Gas Transport

Notes: [1] Suspended

[2] Annual contracted volumes are shown although these are often not being met. Where differences exist the predicted actual figure is added.

Table 44. **TRADE IN LNG 1990**

(million tonnes oil equivalent)

Suppliers	Importers			
	USA	Japan	Western Europe	Total Exports
Abu Dhabi	-	3.2	-	3.2
Algeria	13.8	-	13.0	26.8
Libya	-	-	3.0	3.0
Qatar	-	6.6	-	6.6
OAPEC	13.8	9.8	16.1	39.6
Indonesia	5.1	17.2	-	22.3
Nigeria	6.6	-	6.6	13.2
Other OPEC	11.7	17.2	6.6	35.5
Trinidad	4.4	-	-	4.4
Cameroun	4.4	-	4.4	8.8
USA (Alaska)	3.5	1.2	-	4.7
Canada (Arctic)	2.2	-	-	2.2
Brunei	-	6.6	-	6.6
Malaysia	-	7.7	-	7.7
Australia	-	8.3	-	8.3
Others	14.5	2.4	4.4	42.7
TOTAL IMPORTS	40.0	50.8	27.1	117.9

Source: Own Calculations

Table 45. WORLD GAS TRADE 1979–1990

	1979		1990		% Change 1979–1990
	MTOE	% of Total	MTOE	% of Total	
LNG:					
OPEC	23.0	16	88.5	30	+ 285
Others	8.6	6	32.4	11	+ 277
Total	31.6	22	120.9	41	+ 283
Pipeline:					
Netherlands	38.9	27	26.5	9	– 32
USSR	30.3	21	44.2	15	+ 46
Canada	23.0	16	26.5	9	+ 15
Norway	13.0	9	35.4	12	+ 172
OPEC	2.9	2	20.6	7	+ 610
Mexico)	4.3)	3	11.8	4	n.a.
Others)	)		8.8	3	n.a.
Total	112.4	78	174.1	59	+ 55
GRAND TOTAL	144.0	100	295.0	100	+ 105
OPEC	25.9	18	109.1	37	+ 321
World consumption	1,297.0	-	1,799.0	-	-

Source: Shell International Gas. / Own Estimates

Table 46. **PROVED OIL RESERVES WORLD REGIONS 1970–1979**

(million tonnes of oil)

	1970		1975	1977	1978	1979		
	MT	% of Total	MT	MT	MT	MT	% of Total	% Change 1970-79
Middle East	46,902	56	50,079	49,785	50,236	49,240	56	+ 5
E. Europe, USSR & China	13,854	17	14,066	13,392	12,824	12,259	14	– 11
Africa	9,895	12	8,695	7,890	7,698	7,584	9	– 23
Central and South America	3,589	4	4,972	5,680	7,558	7,956	9	+ 122
North America	6,705	8	5,362	4,783	4,650	4,487	5	– 33
Western Europe	467	1	3,458	3,658	3,263	3,175	4	+ 580
Far East & Australia	1,941	2	2,881	2,648	2,693	2,592	3	+ 33
World TOTAL	83,351	100	89,513	87,836	88,922	87,293	100	+ 5
World Oil Reserves Index (1970 = 100)	100		107	105	107	105		

Table 47(a). OIL-CUMULATIVE ESTIMATED RESOURCES BY REGION (1ST JANUARY 1976)
(Excluding shale oil, tar sands etc.)

(billion tonnes oil)

	Production to end 1975	**Reserves proven and prospective**		**undiscovered**		**Total proven, prospective and un-discovered**	
	BT	**BT**	**% of Total**	**BT**	**% of Total**	**BT**	**% of Total**
USA	16	7	6	11	8	18	7
Other W. Hemisphere	8	12	10	25	18	37	14
Russia, China etc.	8	14	12	50	35	64	25
Middle East	12	68	59	19	13	87	34
Other E. Hemisphere	4	14	12	36	25	50	19
TOTAL	48	115	100	141	100	256	100

Source: M.T. Halbouty and J. D. Moody.

Table 47(b). RECOVERABLE OIL – CUMULATIVE ESTIMATED RESOURCES BY REGION (DECEMBER 1978)

	Billion Tonnes	% Share
Middle East	81.3	29.90
Communist countries	64.2	23.60
USA	29.2	10.75
Africa	22.0	8.10
South America	22.0	8.00
Asia Pacific	13.1	4.80
Canada	11.4	4.20
Western Europe	9.2	3.40
Others (including Mexico)	19.7	7.23
TOTAL	272.0	100.00

Source: M. King Hubbert for the Congressional Research Service, US Senate's Committee on Energy and Natural Resources December 1978.

Table 48. **PROVED OIL RESERVES**

FAR EAST AND AUSTRALIA 1970–1979

(million tonnes of oil)

	1970		1975	1977	1978	1979		
	MT	% of Total	MT	MT	MT	MT	% of Total	% Change 1978-79
Indonesia	1,361	70	1,905	1,361	1,388	1,306	50	– 4
Malaysia	-	-	324	324	363	363	14	†
India	128	7	123	403	390	349	13	+ 173
Australia	262	13	219	257	270	274	11	+ 5
Brunei	130	7	273	211	202	245	9	+ 88
Others	60	3	37	92	80	55	2	– 8
TOTAL	1,941	100	2,881	2,648	2,693	2,592	100	33

Source: Le Petrole en Chiffres / Own Calculations.

Table 49. **PROVED OIL RESERVES – WESTERN EUROPE 1970–1979**

(million tonnes oil)

	1970		1975	1977	1978	1979		
	MT	% of Total	MT	MT	MT	MT	% of Total	% Change 1978-79
U.K.	137	29	2,198	2,610	2,198	2,115	67	+1,444
Norway	135	29	940	806	793	772	24	+ 472
Italy	33	7	103	88	95	95	3	+ 188
West Germany	80	17	72	68	66	66	2	– 17
Denmark	-	-	32	7	39	49	1	-
Spain	1	-	34	36	11	21	1	-
Austria	26	6	24	22	22	20	1	– 23
Netherlands	38	8	37	12	10	9	–	– 76
France	17	4	12	6	8	7	-	– 59
Others	-	-	6	6	21	21	1	-
TOTAL	467	100	3,458	3,658	3,263	3,175	100	+ 580

Source: Le Petrole en Chiffres / Own Calculations.

Table 50. PROVED OIL RESERVES – MIDDLE EAST (INCLUDING TURKEY) 1970-1979

(million tonnes of oil)

	1970		1975	1977	1978	1979		
	MT	% of Total	MT	MT	MT	MT	% of Total	% Change 1970-79
Saudi Arabia	17,276	37	20,251	20,442	22,581	22,261	45	+ 29
Kuwait	9,239	20	9,365	9,227	9,117	9,007	18	– 25
Iran	9,498	20	8,752	8,413	7,916	7,870	16	– 17
Iraq	4,294	9	4,602	4,629	4,307	4,159	8	– 3
Abu Dhabi	1,575	3	3,875	4,073	3,935	3,673	7	+133
Neutral Zone	3,752	8	938	908	949	917	2	– 76
Qatar	558	1.2	759	740	528	497	1.0	– 11
Oman	233	0.5	798	765	338	325	0.7	+ 39
Syria	162	0.3	323	310	300	288	0.6	+ 78
Dubai	135	0.3	179	186	178	192	0.4	+ 42
Bahrain	86	0.2	43	37	34	33	0.1	– 62
Turkey	92	0.2	15	52	50	17	-	– 81
Others	2	-	179	3	3	1	-	-
TOTAL	46,902	100	50,079	49,785	50,236	49,240	100	+ 5

Source: Le Petrole en Chiffres / Own Calculations.

Table 51. PROVED OIL RESERVES – NORTH AMERICA 1970–1979

(million tonnes of oil)

	1970		1975	1977	1978	1979		
	MT	% of Total	MT	MT	MT	MT	% of Total	% Change 1975-79
Canada	1,447	22	956	808	808	915	20	– 37
USA	5,258	78	4,406	3,975	3,842	3,572	80	– 32
TOTAL	6,705	100	5,362	4,783	4,650	4,487	100	– 33

Source: Le Petrole en Chiffres / Own Calculations.

Table 52. PROVED OIL RESERVES – AFRICA 1970–1979

(million tonnes of oil)

	1970		1975	1977	1978	1979		
	MT	% of Total	MT	MT	MT	MT	% of Total	% Change 1978-79
Libya	3,846	39	3,427	3,283	3,191	3,086	41	– 20
Nigeria	1,263	13	2,726	2,524	2,456	2,349	31	+ 86
Algeria	3,890	39	962	862	822	1,102	14	– 72
Egypt	657	7	563	354	442	428	6	– 35
Angola	69	1	180	161	155	167	2	+142
Gabon	96	1	304	283	272	67	1	– 28
Others	72	-	533	423	360	384	5	+433
TOTAL	9,893	100	8,695	7,890	7,698	7,584	100	– 23

Source: Le Petrole en Chiffres / Own Calculations.

Table 53. **PROVED OIL RESERVES EASTERN EUROPE, USSR AND CHINA 1970–1979**

(million tonnes of oil)

	1970		1975	1977	1978	1979		
	MT	% of Total	MT	MT	MT	Mt	% of Total	% Change 1978-79
USSR	10,850	78	10,930	10,204	9,660	9,115	74	– 16
China	2,750	20	2,740	2,740	2,740	2,740	22	-
Rumania	110	1	130	220	210	200	2	+ 82
Hungary	35	-	131	120	110	105	1	+200
Yugoslavia	44	-	51	44	40	40	-	-
Others	65	1	84	64	64	39	-	– 40
TOTAL	13,854	100	14,066	13,392	12,824	12,259	100	– 11

Source: Le Petrole en Chiffres / Own Calculations.

Table 54. **PROVED OIL RESERVES**

CENTRAL AND SOUTH AMERICA 1970–1979

(million tonnes of oil)

	1970		1975	1977	1978	1979		
	MT	% of Total	MT	MT	MT	MT	% of Total	% Change 1978-79
Mexico	450	12	1,337	1,971	4,000	4,400	55	+ 874
Venezuela	1,878	52	2,527	2,598	2,570	2,550	32	+ 36
Argentina	644	18	344	348	334	334	4	– 48
Brazil	116	3	107	120	164	167	2	+ 44
Ecuador	100	3	323	216	154	145	2	+ 45
Colombia	237	7	79	136	106	101	1	– 57
Trinidad	82	2	100	93	72	100	1	+ 22
Peru	36	1	102	97	74	87	1	+ 142
Chile	16	-	24	56	51	51	1	+ 219
Others	30	1	29	45	33	21	-	– 30
TOTAL	3,589	100	4,972	5,680	7,558	7,956	100	+ 122

Source: Le Petrole en Chiffres / Own Calculations.

Table 55. **OIL CONSUMPTION 1979–2000**

(million tonnes of oil)

	1979		1990		2000		
	MT	**% of Total**	**MT**	**% of Total**	**MT**	**% of Total**	**% Change 1979–2000**
USA	862.9	27.7	883[1]	24.7	916[1]	22.3	+ 6
USSR	441.0	14.1	548	15.3	611	14.9	+ 38
China	91.1	2.9	101	2.8	122	3.0	+ 34
Canada	89.9	2.9	112[1]	3.1	120[1]	2.9	+ 3
Western Europe	726.5	23.3	719	20.1	707	17.2	– 3
Japan	265.4	8.5	292	8.2	264	6.4	– 1
Middle East	74.8	2.4	137	3.8	350	8.5	+ 368
Africa	63.5	2.0	124	3.5	260	6.3	+ 309
Australasia	38.0	1.2	41	1.1	38	0.9	-
Latin America[2]	211.8	6.8	284	7.9	346	8.4	+ 63
Eastern Europe	101.1	3.2	114	3.2	117	2.8	+ 16
South Asia	36.7	1.2	46	1.3	59	1.4	+ 61
S.E.Asia	116.9	3.7	170	4.8	199	4.8	+ 70
WORLD	3,119.6	100.0	3,571	100.0	4,109	100.0	+ 32
Communist World	633.2	20.3	763	21.4	850	20.7	+ 34
Free World	2,486.4	27.7	2,808	78.6	3,259	79.3	+ 31

Source: Own Estimates. Notes: [1] Includes synthetics and natural

Table 56. WORLD OIL CONSUMPTION VERSUS RESERVES BY AREA 1979

	Known Reserves (million tonnes)	Consumption (million tonnes per annum)	Years (potential life)
USA	3,572	863	4.1
Canada	915	90	10.2
North America	4,487	953	4.7
Latin America	7,584	212	35.8
Western Hemisphere	16,558	1,165	28.4
Western Europe	3,115	726	4.4
Middle East	49,240	75	656.5
Africa	7,584	63	120.4
USSR	9,115	441	20.7
Eastern Europe	344	101	3.4
China	2,740	91	30.1
Other Eastern Hemisphere	2,600	458	5.7
Total Eastern Hemisphere	70,735	1,955	36.2
World (exc. USSR, Europe and China)	75,094	2,487	30.2
WORLD	87,293	3,120	28.0
Communist World	12,199	633	19.3

Source: BP / Own Calculations

Table 57. WORLD OIL CONSUMPTION BY MAIN CONSUMING COUNTRIES 1969–79
(Unit: million tonnes of oil)

	1969	1970	1971	1972	1973	1974	1975	1976	1977	1978	1979	% of total	% Yearly Change 1974-79
USA	667.8	694.6	719.3	775.8	818.0	782.6	765.9	822.4	865.9	888.8	862.9	27.7	+2.0
USSR	238.6	263.0	279.2	302.9	325.7	358.5	375.1	384.9	399.6	419.2	441.0	14.1	+4.2
Japan	169.0	199.1	219.7	234.4	269.1	258.9	244.0	253.5	260.4	262.7	265.4	8.5	+0.5
W. Germany	117.1	128.6	133.5	140.9	149.7	134.3	1289	138.9	137.1	142.7	146.9	4.7	+1.8
France	83.0	94.3	102.8	114.1	127.3	121.0	110.4	119.5	114.6	119.0	118.1	3.8	–0.5
Italy	77.3	87.3	93.8	98.2	103.6	100.8	94.5	98.8	96.1	99.8	101.2	3.2	+0.1
UK	97.3	103.6	104.3	110.5	113.2	105.3	92.0	91.4	92.0	94.0	94.1	3.0	–2.2
China	20.4	28.2	36.7	43.1	53.8	61.9	68.3	76.9	82.0	84.7	91.1	2.9	+8.0
Canada	69.1	73.0	75.8	79.3	83.7	84.8	83.1	85.9	85.6	86.9	98.9	2.9	+1.2
Spain	24.6	28.1	30.9	32.5	39.1	41.1	42.7	48.3	45.5	47.0	47.3	1.5	+2.9
Netherlands	32.9	36.5	36.0	40.1	41.3	35.4	34.8	39.2	37.6	37.5	38.5	1.2	+1.7
Belg/Lux.	25.1	27.9	28.4	31.1	31.5	28.1	26.5	28.0	28.0	29.0	29.4	0.9	+0.9
Sweden	26.8	29.9	28.2	28.6	29.4	27.1	26.6	29.6	28.1	26.7	28.4	0.9	+0.9
Denmark	16.5	18.4	18.4	19.4	17.9	16.0	15.7	16.7	16.6	16.1	16.1	0.5	+0.1
Yugoslavia	6.0	7.0	9.0	10.1	11.3	11.5	12.2	13.2	13.9	14.8	15.8	0.5	+6.6
Turkey	7.1	7.7	9.0	10.0	12.4	12.5	13.4	15.4	16.6	15.3	15.2	0.5	+4.0
Finland	9.6	10.8	11.1	11.9	13.3	11.6	11.9	12 8	12.5	12.5	13.3	0.4	+2.8
Switzerland	11.2	12.5	13.3	13.6	14.7	13.0	12.5	13.0	13 1	13.4	12.9	0.4	–0.2
Austria	8.3	9.1	10.1	10.9	11.8	10.6	10.7	11.6	11.1	12.0	12.4	0.4	+3.2
Greece	6.2	6.7	7.4	8.6	10.0	9.4	9.9	10.6	10.8	11.7	11.9	0.4	+4 9
Norway	7.4	8.3	8.2	8.5	8.6	7.7	8.0	9.0	8.9	8.7	9.0	0.3	+3.1
Portugal	3.9	4.6	5.4	5.9	6.3	6.6	6.8	7.1	7.1	7.4	7.7	0.2	+3 2
Eire	3.7	4.1	4.5	5.0	5.4	5.4	5.2	5.3	5.7	6.0	6.2	0.2	+3.1
Cyprus/ Gibraltar	1.0	1.0	1.2	1.3	1.4	1.3	1.1	1.3	1.4	1.5	1.5	-	+3.1
World	2,099.8	2,284.0	2,417.1	2,589.4	2,793.4	2,756.9	2,722.8	2,897.3	2,985.6	3,083.2	3,119.6		+2.5

Source: BP.

Table 58. WORLD OIL CONSUMPTION BY REGION[3] 1969–79

(Unit: million tonnes of oil)

	1969	1970	1971	1972	1973	1974	1975	1976	1977	1978	1979	% of total	% Yearly Change 1974–79
North America	736.9	767.6	795.1	855.1	901.7	867.4	849.0	908.3	951.5	975.7	952.8	30.5	+1.9
Latin America	126.8	137.2	147.6	151.5	163.7	171.3	174.0	185.7	193.4	201.4	211.8	6.8	+4.3
Western Hemisphere	863.7	904.8	942.7	1,006.6	1,065.4	1,038.7	1,094.0	1,144.9	1,177.1	1,164.6	37.3	+3.0	
Western Europe [2]	565.5	627.0	656.0	701.8	748.9	699.3	664.4	710.3	697.3	715.7	726.5	23.3	+0.8
Middle East	47.2	49.5	54.0	56.9	62.2	67.1	66.8	74.7	78.9	83.3	74.8	2.4	+2.2
Africa	38.8	42.1	44.4	44.7	49.5	50.4	51.5	55.5	58.0	61.4	63.5	2.0	+4.7
S. Asia	26.5	26.8	27.9	28.9	31.3	29.6	30.1	32.6	34.5	37.1	36.7	1.2	+4.4
S.E. Asia	53.4	59.6	63.9	71.2	77.6	79.2	81.2	88.7	95.8	105.5	116.9	3.7	+8.1
Japan	169.0	199.1	219.7	234.4	269.1	258.9	244.0	253.5	260.4	262.7	265.4	8.5	+0.5
Australasia	27.9	29.7	31.3	31.7	34.8	35.8	35.1	36.5	38.0	37.6	38.0	1.2	+1.2
E. Europe [1]	48.8	54.2	61.3	67.2	75.1	77.5	83.3	89.7	96.2	98.9	101.1	3.2	+5.5
Total E. Hemisphere	1,236.1	1,379.2	1,474.4	1,582.9	1,728.0	1,718.0	1,699.8	1,803.3	1,840.7	1,906.1	1,955.0	62.7	+2.6
Communist Countries[1]	307.8	345.4	377.2	413.2	454.6	497.8	526.7	551.5	577.8	602.8	633.2	20.3	+4.9
World	2,099.8	2,284.0	2,417.1	2,589.4	2,793.4	2,756.9	2,727.8	2,897.3	2,985.6	3,083.2	3,119.6	100	+2.5

1 excluding Yugoslavia

2 including Yugoslavia.

3 including Japan

Source: BP.

Table 59. OIL CONSUMPTION - AFRICA 1970-1979

(million tonnes of oil)

	1970		1975	1977	1978	1979		
	MT	% of Total	MT	MT	MT	MT	% of Total	% Change 1970-79
South Africa	11.2	27	15.2	15.0	15.3	15.5	24	+ 38
Egypt	5.8	14	8.5	11.0	11.2	11.5	18	+ 98
Algeria	2.2	5	3.9	4.4	4.5	4.6	7	+ 109
Libya	1.1	3	2.0	3.4	3.7	3.9	6	+ 254
Nigeria	1.6	4	3.1	3.5	3.5	3.6	6	+ 125
Others	20.2	48	18.8	20.7	23.2	24.4	38	+ 21
TOTAL	42.1	100	51.5	58.0	61.4	63.5	100	+ 51

Source: Le Petrole en Chiffres / Own Calculations

Table 60. OIL CONSUMPTION - FAR EAST AND AUSTRALIA 1970-1979

(million tonnes of oil)

	1970		1975	1977	1978	1979		
	MT	% of Total	MT	MT	MT	MT	% of Total	% Change 1970-79
Japan	199.1	64	244.0	260.4	262.7	265.4	64	+ 33
Australia	25.4	8	28.8	30.8	30.5	30.0	7	+ 18
India	16.2	5	20.9	21.8	22.3	22.0	5	+ 36
Indonesia	7.1	2	15.8	19.5	20.0	20.0	5	+ 182
Singapore	8.1	3	11.4	11.5	11.6	11.4	3	+ 41
Philippines	8.4	3	9.1	10.5	10.5	10.3	2	+ 23
Malaysia	3.6	1	4.0	4.5	4.7	5.0	1	+ 39
Pakistan	4.8	1	3.9	4.0	4.0	3.9	1	- 19
Others	36.0	12	42.2	46.8	46.9	47.4	11	+ 32
TOTAL	308.7	100	380.1	409.8	413.2	415.4	100	+ 35

Source: Le Petrole en Chiffres / Own Calculations

Table 61. **OIL CONSUMPTION - MIDDLE EAST 1970-1979 (INCLUDING TURKEY)**

(million tonnes of oil)

	1970		1975	1977	1978	1979		
	MT	% of Total	MT	MT	MT	MT	% of Total	% Change 1970-79
Iran	16.1	32	24.7	28.0	29.0	27.0	36	+ 68
Saudi Arabia	14.9	30	13.8	17.5	18.0	18.5	25	+ 24
Israel	5.4	11	7.0	7.0	7.3	7.5	10	+ 39
Iraq	3.6	7	4.1	4.2	4.3	4.5	6	+ 25
Kuwait	4.7	9	3.6	3.9	4.0	4.0	5	- 15
Others	4.8	10	13.6	18.3	20.7	13.3	18	+ 177
TOTAL	49.5	100	66.8	78.9	83.3	74.8	100	+ 51

Source: Le Petrole en Chiffres / Own Calculations

Table 62. OIL CONSUMPTION - EASTERN EUROPE, USSR AND CHINA 1970-1979

(million tonnes of oil)

	1970		1975	1977	1978	1979		
	MT	% of Total	MT	MT	MT	MT	% of Total	% Change 1970-79
USSR	26.3	75	375.1	399.8	419.2	441.0	68	+ 68
China	28.2	8	68.3	82.0	84.7	91.1	14	+ 223
Czechoslovakia	9.5	3	15.9	19.3	20.0	20.3	3	+ 114
Rumania	10.0	3	13.9	17.5	19.0	19.0	3	+ 90
East Germany	9.8	3	14.7	17.0	18.0	18.5	3	+ 89
Poland	8.5	2	14.5	17.5	17.0	16.5	2	+ 94
Yugoslavia	7.0	2	12.2	13.9	14.7	15.8	2	+ 126
Bulgaria	8.2	2	13.0	13.5	13.8	13.9	2	+ 69
Hungary	6.5	2	10.1	10.7	11.0	12.0	2	+ 85
Others	1.9	-	0.2	0.7	0.7	0.9	-	- 53
TOTAL	352.6	100	538.9	591.7	617.6	649	100	+ 84

Source: Le Petrole en Chiffres / Own Calculations

Table 63. **OIL CONSUMPTION - CENTRAL AND SOUTH AMERICA 1970-1979**

(million tonnes of oil)

	1970		1975	1977	1978	1979		
	MT	% of Total	MT	MT	MT	MT	% of Total	% Change 1970-79
Brazil	25.2	18.4	42.7	46.5	51.0	35.0	16.5	+ 39
Mexico	24.1	17.6	33.3	38.0	39.0	41.0	19.3	+ 70
Argentina	22.1	16.1	22.5	24.0	23.4	25.0	11.8	+ 13
Venezuela	13.5	9.8	12.0	13.0	14.8	15.5	7.3	+ 15
Cuba	6.6	4.8	7.9	8.4	8.5	8.5	4.0	+ 29
Puerto Rico	6.9	5.0	7.3	7.8	7.8	7.6	3.6	+ 10
Colombia	5.7	4.1	6.4	7.1	7.1	7.0	3.3	+ 23
Antilles	5.2	3.8	6.0	6.6	6.6	6.5	3.1	+ 25
Peru	5.3	3.9	6.4	6.4	6.4	6.2	2.9	+ 17
Chile	4.8	3.5	4.6	4.9	4.9	4.7	2.2	- 2
Trinidad	3.7	2.7	2.7	3.0	3.0	3.0	1.4	- 19
Others	14.1	10.3	22.2	27.7	28.9	31.8	15.0	+ 9
TOTAL	137.2	100.0	174.0	193.4	201.4	211.8	100.0	+ 54

Source: Le Petrole en Chiffres / Own Calculations

Table 64.

WORLD OIL PRODUCTION 1969-2000

(Unit: million tonnes of oil)

	USA	USSR	China	E.Europe	Canada	W.Europe	Middle East	Africa	Australasia	Latin America	S.Asia	S.E.Asia	World
1969	458.7	328.4	20.4	17.3	62.2	23.6	615.4	249.2	2.0	264.9	7.5	43.9	2,150.6
1970	478.6	353.0	28.2	17.6	71.5	22.8	691.7	299.9	8.6	272.7	8.2	50.0	2,362.5
1971	669.9	377.1	36.7	18.2	76.6	21.9	807.7	281.1	15.0	266.2	8.6	55.1	2,494.9
1972	470.1	400.4	42.1	19.3	88.8	22.3	898.6	282.1	16.8	259.6	9.2	65.7	2,633.8
1973	457.3	429.0	54.8	19.2	102.3	22.6	1,052.5	290.0	18.5	272.1	9.0	82.0	2,871.7
1974	436.8	458.9	65.8	19.7	96.5	22.6	1,085.0	269.4	18.4	254.7	8.9	81.9	2,879.2
1975	415.9	490.8	74.3	20.0	83.5	30.8	973.1	248.5	19.9	227.6	9.6	78.5	2,733.1
1976	404.9	519.7	83.6	20.0	77.2	45.3	1,107.1	285.8	20.5	229.5	10.0	92.5	2,953.8
1977	409.4	545.8	93.6	20.0	75.7	69.7	1,118.4	305.2	21.4	238.6	11.9	103.8	3,071.5
1978	432.4	572.5	104.1	21.0	74.4	89.6	1,058.9	297.1	21.4	251.5	12.6	103.4	3,095.4
1979	423.7	586.0	106.1	20.0	86.0	115.9	1,076.4	324.4	21.9	280.0	14.9	106.5	3,221.7
% of total	13.1	18.2	3	0.6	3.3	3.6	33.4	10.1	0.7	8.7	0.5	3.3	100
% change 1969–79	–8	+78	+420	+16	+38	+391	+75	+30	+995	+6	+99	+143	+50
1990	413*	646	137	21	117	175	1,289	335	22	390	25	118	3,688
% of total	11.2	17.5	4	0.6	3.2	4.7	34.9	9.1	0.6	10.6	2.7	3.2	100
% change 1979–1990	–2	+10	+29	+5	+36	+51	+20	+2	-	+39	+68	+11	+14
2000	546*	687	162	21	130	142	1,470	350	30	530	41	130	4,243
% of total	12.9	16.2	3.8	0.5	3.1	3.3	34.6	8.2	0.7	12.5	1.0	3.1	100
% change 1979-2000	+29	+17	+53	+5	+51	+22	+37	+8	+37	+89	+175	+22	+32

* includes synthetics.

Source BP/Own Calculations.

Table 65. WORLD OIL PRODUCTION BY MAIN PRODUCERS 1969-1979

(Unit: million tonnes of oil)

	% of total	1969	1970	1971	1972	1973	1974	1975	1976	1977	1978	1979	% of total	% Change 1978–1979
USSR	15.3	328.4	353.0	377.1	400.4	429.0	458.9	490.8	519.7	545.8	572.5	586.0	18.2	+ 2.4
Saudi Arabia †	7.0	150.2	178.0	225.0	287.2	367.9	412.4	343.9	421.6	455.0	409.8	468.3	14.5	+14.3
United States	21.3	458.7	478.6	469.9	470.1	457.3	436.8	415.9	404.9	409.4	432.4	423.7	13.1	– 2.0
Iraq †	3.5	74.9	76.9	83.5	72.1	99.0	96.7	111.0	118.8	122.3	127.6	169.3	5.2	+32.7
Iran †	7.8	168.1	191.3	227.0	251.9	293.2	301.2	267.7	295.0	283.5	260.4	155.6	4.8	– 4.0
Nigeria †	1.2	26.4	52.9	74.7	88.9	100.1	112.2	88.8	102.9	104.1	95.1	114.2	3.5	+ 2.0
Venezuela †	8.8	188.7	195.2	187.7	171.5	179.0	158.5	125.3	122.9	119.5	115.4	125.4	4.0	+ 8.7
Kuwait †	6.1	131.0	139.1	148.8	153.0	140.4	116.3	94.0	98.2	91.5	97.0	114.1	3.5	+17.6
China	0.9	20.4	28.2	36.7	42.1	54.8	65.8	74.3	83.6	93.6	104.1	106.1	3.3	+ 1.9
Libya †	7.0	149.9	159.8	133.1	108.2	104.9	73.3	71.3	93.3	99.4	95.2	99.6	3.1	+ 4.6
Indonesia †	1.7	37.1	42.2	44.1	53.4	66.0	67.9	64.6	74.6	83.5	81.0	78.8	2.4	– 2.7
UK	-	0.1	0.1	0.1	0.1	0.1	0.1	1.4	11.8	37.5	53.3	77.9	2.4	+ 4.6
Mexico	1.1	22.8	23.9	23.8	24.8	26.9	31.6	39.3	43.6	53.7	66.0	80.5	2.5	+22.0
Canada	3.0	62.2	71.5	76.6	88.8	102.3	96.5	83.5	77.2	75.7	74.4	86.0	2.7	+15.6
Abu Dhabi †	1.3	28.9	33.4	44.9	50.6	62.6	67.7	67.3	76.8	80.0	69.7	70.2	2.2	+ 0.7
Algeria †	2.1	44.5	48.5	36.5	49.8	51.2	47.1	47.5	50.1	53.5	57.2	56.4	1.7	– 1.4
Egypt	0.8	17.1	23.5	21.0	17.6	13.0	11.5	14.8	16.4	21.0	24.2	25.5	0.8	+ 5.4
Qatar †	0.8	17.0	17.7	20.5	23.2	27.3	24.9	21.0	23.4	21.1	23.4	24.6	0.8	+ 5.1
Norway	-	-	-	0.3	1.6	1.8	1.7	9.3	13.8	13.5	17.2	18.8	0.6	+ 9.3
Dubai †	-	0.5	4.3	6.2	7.6	10.8	12.0	12.6	15.6	15.8	18.0	17.6	0.5	– 2.2
North America	26.8	577.3	609.0	606.6	621.0	621.3	593.2	557.4	539.2	542.5	562.5	569.1	17.7	+ 1.2
Latin America	12.3	264.9	272.7	262.2	255.6	272.1	254.7	227.6	229.5	238.6	251.5	280.0	8.7	+11.3
W. Europe *	1.1	23.6	22.8	21.9	22.3	22.6	22.6	30.8	45.3	69.7	89.6	115.9	3.6	+29.3
Middle East	28.6	615.4	691.7	807.7	898.6	1,052.5	1,085.0	975.1	1,107.1	1,118.4	1,058.9	1,076.4	33.4	+ 1.6
Africa	11.6	249.2	299.9	281.1	282.2	290.0	269.4	248.5	285.8	305.2	297.1	324.4	10.1	+ 9.2
S.E. Asia	2.0	43.9	50.0	55.1	65.7	82.0	81.9	78.5	92.5	103.8	103.4	106.5	3.3	+ 3.0
World (non-Communist)	83.0	1,784.5	1,963.7	2,062.9	2,172.0	2,368.7	2,334.8	2,148.0	2,330.5	2,412.1	2,397.8	2,509.6	77.9	+ 4.7
Communist World	17.9	386.1	398.8	432.	461.8	503.0	544.4	585.1	623.3	659.4	697.6	712.1	22.1	+ 2.1
WORLD	100	2,150.6	2,362.5	2,494.9	2,633.8	2,871.7	2,879.2	2,733.1	2,953.8	3,071.5	3,095.4	3,221.7	100	+ 4.1

*excluding Turkey † OPEC member.

Source: BP / Own Calculations.

Table 66. **WORLD OIL TRADE – 1979**

(Unit: million tonnes of oil)

FROM \ TO	USA	CANADA	LATIN AMERICA	WESTERN EUROPE	AFRICA	SOUTH E. ASIA	JAPAN	AUSTRA-LASIA	Other E. Hemisphere	Destination not known	TOTAL EXPORTS	% of Total
USA	-	5.3	12.1	5.6	-	0.8	1.8	0.2	-	-	25.8	1.5
Canada	22.6	-	-	-	-	-	-	-	-	-	22.6	1.3
Latin America	125.5	11.9	11.5	14.5	1.2	-	0.5	-	7.2	16.3	188.6	10.8
W. Europe	17.5	-	-	-	7.1	-	-	-	1.2	-	25.8	1.5
Middle East	104.6	12.5	74.5	430.6	22.2	86.7	205.2	14.0	44.7	14.6	1,009.6	57.6
N. Africa	64.8	1.0	6.0	89.1	1.0	-	0.7	-	1.2	-	163.8	9.3
W. Africa	58.9	-	19.5	50.0	3.0	-	-	-	-	-	131.4	7.5
E. & S. Africa	-	-	-	-	-	0.6	-	-	-	-	0.6	-
South Asia	-	-	-	-	-	0.5	0.5	-	1.1	-	2.1	0.1
South East Asia	25.5	-	1.2	1.2	0.5	-	58.2	4.4	1.0	-	92.0	5.2
Japan	-	-	-	-	-	0.4	-	-	-	-	0.4	-
Australasia	0.4	-	-	-	-	-	0.7	-	-	-	1.1	0.1
USSR, E. Europe & China	-	-	10.0	56.1	1.7	9.0	8.0	-	3.0	-	87.8	5.0
Total IMPORTS	419.8	30.7	134.8	647.1	36.7	98.0	275.6	18.6	59.4	30.9	1,751.6	100
(% of Total)	24.0	1.7	7.7	36.9	2.1	5.6	15.7	1.1	3.4	1.8	100	

Source: BP.

Table 67. O.P.E.C. OIL PRODUCTION 1970–79

(Unit: Million tonnes of oil)

	1970		1975	1977	1978	1979		% change 1970–79
	MT	% of Total	MT	MT	MT	MT	% of Total	
Venezuela	195.2	16.7	125.3	119.5	115.4	125.4	8.1	– 36
Ecuador	0.2	-	7.9	9.1	10.0	10.5	0.7	-
Iran	191.3	16.3	267.7	283.5	260.4	155.6	10.1	– 19
Iraq	76.9	6.6	111.0	122.3	127.6	169.3	11.0	+ 120
Kuwait	139.1	11.9	94.0	91.5	97.0	114.1	7.4	– 18
Qatar	17.7	1.5	21.0	21.0	23.4	24.6	1.6	+ 39
Neutral Zone	26.0	2.2	25.8	18.5	23.9	29.0	1.9	+ 11
Saudi Arabia	178.0	15.2	343.9	455.0	409.8	468.3	30.3	+ 163
Abu Dhabi	33.4	2.8	67.3	80.0	69.7	70.2	4.5	+ 110
Sharjah	-	-	1.9	1.4	1.1	0.7	-	-
Dubai	4.3	0.4	12.6	15.8	18.0	17.6	1.1	+ 309
Algeria	48.5	4.1	47.5	53.5	57.2	56.4	3.6	+ 16

contd......

Table 67. continued ....

	1970		1975	1977	1978	1979		% change
	MT	% of Total	MT	MT	MT	MT	% of Total	1970–79
Libya	159.8	13.6	71.3	99.4	95.2	99.6	6.4	– 38
Gabon	5.4	0.5	11.2	11.1	10.8	10.2	0.7	+ 89
Nigeria	52.9	4.5	88.8	104.1	95.1	114.2	7.4	+ 116
Indonesia	42.2	3.6	64.6	83.5	81.0	78.8	5.1	+ 87
OPEC TOTAL	1,170.9	100	1,361.8	1,569.3	1,495.6	1,544.5	100	+ 32
OPEC Total	1,170.9	49.6	1,361.8	1,569.3	1,495.6	1,544.5	47.9	+ 32
Non-Opec Total	1,191.6	50.4	1,371.3	1,502.2	1,599.8	1,677.2	52.1	+ 41
World Total	2,362.5	100	2,733.1	3,071.5	3,095.4	3,221.7	100	+ 36

Source: Own Calculations / BP / Le Petrole en Chiffres

Table 68. IMPORTS AND EXPORTS 1969–1979

(Unit: million tonnes of oil)

IMPORTS	% of total	1969	1970	1971	1972	1973	1974	1975	1976	1977	1978	1979	% of total	Average Yearly Change 1974–1979 %
USA	14.1	163	176	201	242	316	315	301	365	432	409	420	24.0	+ 5.9
W. Europe	48.4	362	636	667	696	760	734	626	682	657	648	647	36.9	– 2.5
Japan	15.7	182	213	235	238	270	268	245	262	271	263	276	15.7	+ 0.6
Rest of World	21.7	252	245	297	326	348	339	336	397	364	371	409	23.3	+ 3.9
WORLD IMPORTS	100	1,159	1,270	1,400	1,502	1,694	1,656	1,508	1,706	1,724	1 691	1,752	100	
EXPORTS														
USA	1.0	12	14	12	12	13	12	11	12	14	20	26	1.5	+16.4
Canada	2.5	29	36	41	55	68	53	40	31	28	23	23	1.3	–15.5
Latin America	15.9	184	175	190	187	202	182	159	173	162	177	189	10.8	+ 0.7
Middle East	49.3	571	641	756	842	990	992	918	1,033	1,024	975	1.009	57.6	+ 0.3
N. Africa	17.5	203	225	186	167	163	125	119	147	157	161	164	9.4	+ 5.5
W. Africa	3.2	37	54	80	94	99	125	98	112	118	105	131	7.5	+ 1.1
S.E. Asia	3.5	41	40	46	56	67	67	72	84	92	86	92	5.2	+ 6.5
USSR, E. Europe & China	5.0	58	59	66	67	71	73	74	93	101	112	88	5.0	+ 3.8
Other E. Hemisphere	2.1	24	26	23	22	21	27	17	21	28	32	30	1.7	+ 2.4
WORLD EXPORTS	100	1,159	1,270	1,400	1,502	1,694	1,656	1,508	1,706	1.724	1,691	1,752	100.0	

Source: BP.

Table 69. OIL PRODUCTION – MIDDLE EAST (INCLUDING TURKEY) 1970–79

(Unit: Million tonnes of oil)

	1970		1975	1977	1978	1979		% change 1970–79
	M.T.	% of Total	M.T.	M.T.	M.T.	M.T.	% of Total	
Saudi Arabia	178.0	25.6	343.9	455.0	409.8	468.3	43.4	+ 163
Iraq	76.9	11.1	111.0	122.3	127.6	169.3	15.7	+ 120
Iran	191.3	27.5	267.7	283.5	260.4	155.6	14.4	– 19
Kuwait	139.1	20.0	94.0	91.5	97.0	114.1	10.6	– 18
Abu Dhabi	33.4	4.8	67.3	80.0	69.7	70.2	6.5	+ 110
Qatar	17.7	2.5	21.0	21.1	23.4	24.6	2.3	+ 39
Neutral Zone	26.0	3.7	25.8	18.5	23.9	29.0	2.7	+ 11
Dubai	4.3	0.6	12.6	15.8	18.0	17.6	1.6	+ 309
Oman	16.6	2.4	12.1	17.1	15.8	14.8	1.4	– 11
Syria	4.3	0.6	9.6	10.6	10.0	8.5	0.8	+ 91
Turkey	3.5	0.5	3.1	2.7	2.7	2.6	0.2	– 26
Bahrain	3.8	0.5	3.1	2.8	2.7	2.5	0.2	– 34
Others	0.3	-	2.0	1.6	0.5	1.9	0.2	+ 533
TOTAL	695.2	100	978.2	1,121.1	1,061.6	1,079.0	100	+ 55

Source: Le Petrole en chiffres / Euromonitor.

Table 70. **OIL PRODUCTION - AFRICA 1970-1979**

(million tonnes of oil)

	1970		1975	1977	1978	1979		
	MT	% of Total	MT	MT	MT	MT	% of Total	% Change 1970-79
Nigeria	529	18	71.3	104.1	95.1	114.2	35	+ 116
Libya	159.8	53	71.3	91.4	95.2	99.6	31	- 38
Algeria	48.5	16	47.5	53.5	57.2	56.4	17	+ 16
Egypt	23.5	8	14.8	21.0	24.2	25.5	8	+ 8
Gabon	5.4	2	11.2	11.1	10.8	10.2	3	+ 89
Angola	5.1	2	8.4	8.6	8.2	9.0	3	+ 76
Others	4.7	2	6.9	7.5	6.4	9.5	3	+ 102
TOTAL	299.9	100	248.5	305.2	297.1	324.4	100	+ 8

Source: Le Petrole en Chiffres / Own Calculations

Table 71. COMMUNIST WORLD OIL PRODUCTION – EASTERN EUROPE (INCLUDING YUGOSLAVIA) USSR AND CHINA 1970–79

(Unit: Million tonnes of oil equivalent)

	1970		1975	1977	1978	1979		% change
		% of total					% of total	1970–79
USSR	353.0	88	490.8	545.8	572.5	586.0	82	+166
China	28.2	7	74.3	93.6	104.1	106.1	15	+276
Rumania	13.4	3	14.6	14.7	13.7	13.5	2	+ 1
Yugoslavia	2.9	1	3.9	4.0	4.1	4.1	1	+ 41
Hungary	1.9	-	2.0	2.2	2.2	2.0	-	+ 5
Others	2.6	1	3.4	3.1	5.1	4.5	1	+ 73
TOTAL	402	100	589	663.4	701.7	716.2	100	+ 78

Source: Le petrole en chiffres / Own Calculations.

Table 72. **FAR EAST AND AUSTRALIA OIL PRODUCTION 1970–79**

(Unit: Million tonnes of oil)

	1970		1975	1977	1978	1979		% change
	M.T.	% of total	M.T.	M.T.	M.T.	M.T.	% of total	1970–79
Indonesia	42.2	63	64.6	83.5	81.0	78.8	55	+ 87
Australia	8.6	13	19.9	21.4	21.4	21.9	15	+155
India	6.8	10	8.1	10.1	11.0	13.0	9	+ 91
Malaysia	-	-	4.7	9.2	10.8	13.0	9	-
Brunei	6.9	10	9.4	10.9	10.1	11.9	8	+ 72
Others	2.0	3	1.9	2.6	3.2	4.4	3	+120
TOTAL	66.5	100	108.8	137.7	137.5	143.0	100	+115

Source: Le petrole en chiffres / Own Calculations.

Table 73. **OIL PRODUCTION - LATIN AMERICA 1970-1979**

(million tonnes of oil)

	1970		1975	1977	1978	1979		
	MT	% of Total	MT	MT	MT	MT	% of Total	% Change 1970-79
Venezuela	195.2	72	125.3	119.5	115.4	125.4	45	- 36
Mexico	23.9	9	39.3	53.7	66.0	80.5	29	+ 237
Argentina	20.0	7	20.2	21.8	23.0	24.5	9	+ 22
Trinidad	7.2	3	11.1	11.8	11.9	11.5	4	+ 60
Ecuador	0.2	-	7.8	8.7	9.7	10.8	4	-
Peru	3.4	1	3.7	3.9	7.7	9.8	3	+ 188
Brazil	8.0	3	9.4	8.1	8.3	8.5	3	+ 6
Colombia	11.1	4	8.1	7.1	6.8	6.5	2	- 41
Chile	1.6	1	1.1	1.0	0.8	1.1	-	- 31
Others	2.1	1	1.6	3.0	1.9	1.4	-	- 33
TOTAL	272.7	100	227.6	238.6	251.5	280.0	100	+ 3

Source: Le Petrole en Chiffres / Own Calculations

Table 74. **OIL PRODUCTION - WESTERN EUROPE 1970-1979 (NOT INCLUDING YUGOSLAVIA AND TURKEY)**

(million tonnes of oil)

	1970		1975	1977	1978	1979		
	MT	% of Total	MT	MT	MT	MT	% of Total	% Change 1970-79
U.K.	0.1	1	1.4	37.5	53.3	77.9	71	-
Norway	-	-	9.3	13.5	17.2	18.8	17	-
West Germany	7.5	46	5.7	5.4	5.1	4.8	4	- 36
Italy	1.6	10	1.0	1.1	1.5	1.8	2	+ 12
Austria	2.8	17	2.0	1.8	1.8	1.7	2	- 39
France	2.3	14	1.0	1.0	1.1	1.2	1	- 48
Netherlands	1.9	12	1.6	1.6	1.5	1.6	1	- 16
Others	0.2	1	1.8	1.1	1.3	1.4	1	+ 600
TOTAL	16.4	100	23.8	63.0	82.8	109.2	100	+ 566

Source: Le Petrole en Chiffres / Own Calculations

Table 75. **THE WORLD OIL MARKET 1979–2000**

(Unit: million tonnes of oil)

	USA	USSR	China	Japan	Eastern Europe	Canada	Western Europe	Middle East	Africa	Australasia	Latin America	S.Asia	S.E.Asia	World	of which OPEC	%
1979																
Production	423.7	586.0	106.1	0.5	20.0	86.0	115.9	1,076.4	324.4	21.9	280.0	14.9	106.5	3,221.7	1,544.5	47.9
Consumption	862.9	441.0	91.1	265.4	101.1	89.9	726.5	74.8	63.5	38.0	211.8	36.7	116.9	3,119.6	114.5	3.7
Oil Balance	−439.2	+145	+15	−264.1	−81.1	−3.9	−611	+1,001.6	+260.9	−16.1	+68.2	−21.8	−10.4	Total[2] Oil Movement 2,939.1	+1,430	49
1990																
Production	413[1]	646	137	0.5	21	117[1]	175	1,289	335	22	390	25	118	3,688	1,837	49.8
Consumption	883	548	101	292	114	112	719	137	124	41	284	46	170	3,571	261.7	7.3
Oil Balance	−470	+98	+36	−291.5	−93	+5	−544	+1,152	+211	−19	+106	−21	−52	Total[2] Oil Movement 3,098.5	+1,575	51
2000																
Production	546[1]	687	162	0.5	21	130[1]	142	1,470	350	30[1]	530[1]	41	130	4,243	2,094	49.2
Consumption	916	611	122	264	117	120	707	350	260	38	346	59	199	4,109	681.9	16.6
Oil Balance	−370	+76	+40	−263.5	−96	+10	−565	+1,120	+90	−8	+184	−18	−69	Total[2] oil Movement 2,909	+1,412	48

1 includes synthetics
2 does not include trade between countries within regions.

Source: Own Calculations

Table 76. NUCLEAR ENERGY - CONSUMPTION 1969–1979

(Unit: million tonnes of oil equivalent)

	1969	1970	1971	1972	1973	1974	1975	1976	1977	1978	1979	% of Total
USA	3.7	5.7	3.9	15.2	21.8	28.8	44.4	51.6	67.8	75.5	72.2	46
Japan	0.3	1.2	2.1	2.2	2.3	4.6	5.3	9.0	6.4	12.5	14.7	9
USSR	0.8	0.9	1.1	2.1	3.0	4.5	6.0	8.0	10.7	11.0	12.5	7
France	1.2	1.4	2.3	3.6	3.0	3.0	3.9	4.1	5.1	6.4	9.6	6
W. Germany	1.2	1.5	1.4	2.2	2.8	2.9	5.0	5.6	8.3	8.1	9.6	6
Canada	-	-	1.0	1.7	3.7	3.6	3.0	4.1	6.9	8.5	8.5	5
UK	7.5	6.7	7.1	6.5	5.9	7.1	6.3	7.6	8.4	7.9	8.1	5
Eastern Europe	0.1	0.1	0.2	0.5	1.0	1.5	1.4	2.8	2.9	3.6	3.8	3
Sweden	†	†	†	0.4	0.5	0.5	3.0	3.9	4.9	4.1	3.6	2
Belgium/Luxembourg	†	†	†	†	†	†	1.6	2.2	2.6	2.7	2.5	2
Switzerland	0.1	0.6	0.7	0.9	1.6	1.6	1.9	2.0	2.1	2.1	2.4	2
Spain	0.2	0.2	0.7	1.2	1.7	1.9	1.9	1.9	1.7	2.0	1.7	1
Finland	-	-	-	-	-	-	-	-	0.6	0.8	1.6	1
Italy	0.4	0.8	0.9	1.0	0.8	0.8	0.9	1.0	0.9	1.1	1.3	1
S.E. Asia	-	-	-	-	-	-	-	-	-	1.3	1.4	1
Netherlands	0.1	0.1	0.1	0.1	0.3	0.8	0.8	0.9	1.0	0.9	0.9	1
S. Asia	0.4	0.6	0.5	0.8	1.0	0.8	1.0	1.0	0.8	0.6	0.6	1
Latin America	-	-	-	-	-	0.2	0.7	0.7	0.4	0.7	0.8	1
World	15.1	18.8	26.7	35.8	45.4	56.6	79.7	95.6	118.4	135.2	155.8	100

† less than 0.05 T.O.E.

Source: BP.

Table 77. NUCLEAR ENERGY CONSUMPTION 1979-2000

(million tonnes of oil equivalent)

	1979		1990		2000		
	MTOE	% of Total	MTOE	% of Total	MTOE	% of Total	% Increase 1979-2000
USA	72.2	46.0	174	34	286	32	+ 296
USSR	12.5	8.0	29	6	47	5	+ 276
Eastern Europe	3.8	2.0	9	2	17	2	+ 347
China	-	-	-	-	-	-	-
Canada	8.5	5.0	21	4	34	4	+ 300
Western Europe	41.3	26.0	181	36	308	35	+ 646
Japan	14.7	9.0	44	9	105	12	+ 614
Middle East	-						
Africa	-						
Australasia	-						
Latin America	0.8	0.5	48	9	85	10	
South Asia	0.6	0.4					
SE Asia	1.4	0.9					
WORLD	155.8	100.0	506	100	882	100	+ 466

Note: the 1990 and 2000 values 48/9 and 85/10 are bracketed together for Middle East through SE Asia.

Source: Own Calculations

Table 78. **WORLD NUCLEAR CAPACITY BY COUNTRY 1978–1979**

(giga watts)

	MW Installed Year end 1978	MW on Order Year end 1978	Potential Nuclear Capacity (Jan 1979)	% of Total 1978	% of Total Future
USA	52,600	137,000	189,600	48	100
Japan	11,200	7,000	18,200	10	5
West Germany	9,000	18,000	27,000	8	8
France	8,300	36,000	44,300	7	12
Britain	6,000	6,000	14,000	7	4
Sweden	5,400	4,000	9,400	5	3
Canada	5,400	10,000	15,400	5	4
Spain	2,000	12,000	14,000	2	4
USSR	8,500	12,500	21,000	8	6
TOTAL	110,400	242,500	352,900	100	100

Source: Own Calculations.
Various Sources

Table 79. WORLD NUCLEAR POWER 1978–2000

(Unit: million tonnes of oil equivalent and gigawatts)

	1978		1990		2000	
	Year end Capacity GW	MTOE	GW	MTOE	GW	MTOE
United States	52.6	75.5	130	174	212	286
Canada	5.4	8.5	15.4	21	25	34
France	8.3	6.4	44.3	59	66	89
W. Germany	9.0	8.1	27.0	36	44	59
UK	8.0	7.9	14.0	19	28	38
Spain	2.0	2.0	14.0	19	26	35
Italy	0.8	1.1	8.0	11	19	26
Sweden	5.4	4.1	9.4	13	9	11
Other W.Europe	2.7	6.5	18	24	37	50
Total W. Europe	30	36.1	134.7	181	229	308
Japan	11.2	12.5	35.0	44	78	105
Other Free World	12.0	11.1	36.0	48	63	85
Total Free World	111.2	135.2	334	448	607	818
E. Europe	3.9	3.6	7.7	9	13	17
USSR	8.5	11.0	21	29	35	47
WORLD Total	123.6	149.8	379.8	506	655	882

Source: Own Calculations.

Table 80. WORLD[1] URANIUM RESOURCES

(Unit: '000 tonnes of uranium)

	Uranium resources up to $130/KgU				Reasonably assured		Reasonably assured	
	Reasonably assured		Reasonably assured plus estimated additional resources		Uranium resources up to $80/KgU		Uranium resources up to $80/$130/KgU	
	000 tonnes	% of total	000 tonnes	% of total	000 tonnes	% of total	000 tonnes	% of total
United States	708	27	(1,158) 1,886	37	531	29	177	24
South Africa	391	15	(139) 406	8	247	13	144	19
Sweden[3]	301	12	(3) 304	6	0	-	301	41
Australia	299	11	(53) 352	7	290	16	9	1
Canada[2]	235	9	(728) 963	19	215	12	220	30
Niger	160	6	(53) 213	4	160	9	0	-
Namibia	133	5	(53) 186	4	117	6	16	2
Brazil	74.2	3	(90.1) 164.3	3	74.2	4	0	-
France	55.3	2	(46.2) 101.5	2	39.6	2	15 7	2
Gabon	37	1	(0) 37	1	37	2	0	-
India	29.8	1	(23.7) 53.5	1	29.8	2	0	-
Argentina	28.1	1	(9.1) 37.2	1	23	1	5.1	1
Algeria	28	1	(5.5) 33.5	1	28	1	0	-
Denmark	27	1	(16) 53	1	0	-	27	4
Central African Republic	18	1	(0) 18	-	18	1	0	-
Spain	9.8	-	(8.5) 18.3	-	9.8	-	0	-
Portugal	8.2	-	(2.5) 10.7	-	6.7	-	1.5	-
Japan	7.7	-	(0) 7.7	-	7.7	-	0	-
Somalia	6.6	-	(3.4) 10.0	-	0	-	6.6	1
Yugoslavia	6.5	-	(20.5) 27.0	-	4.5	-	2	-

contd......

(Table 80. continued)

	000 tonnes	% of total	000 tonnes	% of total	000 tonnes	% of total	000 tonnes	% of total
Mexico	6	-	(2.4) 8.4	-	6	-	0	-
W. Germany	4.5	-	(7.5) 12.0	-	4	-	0.5	-
Korea (Rep. of)	4.4	-	(0) 4.4	-	0	-	4.4	-
Turkey	3.9	-	(0) 3.9	-	2.4	-	1.5	-
Finland	2.7	-	(0.5) 3.2	-	0	-	0	-
Austria	1.8	-	(0) 1.8	-	1.8	-	0	-
Zaire	1.8	-	(1.7) 3.5	-	1.8	-	0	-
Italy	1.2	-	(2) 3.2	-	0	-	1.2	-
Philippines	0.3	-	(0) 0.3	-	0.3	-	0	-
UK	0	-	(7.4) 7.4	-	0	-	0	-
Chile	0	-	(5.1) 5.1	-	0	-	0	-
Egypt	0	-	(5) 5	-	0	-	0	-
Botswana	0.4	-	(0) 0.4	-	0	-	0.4	-
Madagascar	0	-	(2) 2	-	0	-	0	-
Bolivia	0	-	(0.5) 0.5	-	-	-	0	-
TOTAL	2,590	100	(2,450) 5,040	100	1,850	100	740	100

NOTES

1 Excluding countries with centrally planned economies

2 Reserves mineable at prices up to $ CAN 123/KgU and other reasonably assured resources are mineable at prices between $ CAN 125 and $ CAN 175/KgU.

3 No uranium production allowed in a deposit of below 3000,000 tonnes U. due to a veto by the local authorities on environmental grounds

Source: 'Atom' (UKAEA) / Euromonitor.

Table 81. **EXPORTS OF NUCLEAR REACTORS**

Suppliers	Reactors				Purchasing Countries
	Number	Power MW	% of total	Type	
WESTINGHOUSE (United States)	39	24,326	36	PWR	Belgium, Brazil, South Korea, Spain, Italy, Japan, Philippines, Sweden, Switzerland, Taiwan, Yugoslavia.
of which in service	(15)	(5,155)	(31)		
GENERAL ELECTRIC (United States)	23	12,909	19	BWR	West Germany, Spain, India, Italy, Japan, Mexico, The Netherlands, Switzerland.
of which in service	(14)	(4,596)	(27)		
TECHNOPROMEXPORT (Soviet Union)	37	16,331	24	VVER	East Germany, Bulgaria, Finland, Hungary, Libya, Poland, Czechoslovakia.
of which in service	(8)	(3,158)	(19)		
KWU (West Germany)	8	7,253	11	PWR	Argentina, Austria, Brazil, Spain, The Netherlands, Switzerland.
of which in service	(4)	(1,966)	(12)	PHWR BWR	
AECL (Canada)	7	3,278	5	PHWR	Argentina, South Korea, India, Pakistan, Romania.
of which in service	(2)	(360)	(2)		

contd........

Table 81. continued ......

Suppliers	Reactors				Purchasing Countries
	Number	Power MW	% of total	Type	
FRAMATOME (France)	2	1,914	3	PWR	South Africa.
FRENCH INDUSTRY	(1)	(500)	1 (3)	UNGG	Spain.
BRITISH INDUSTRY	(2)	(376)	- (2)	AGR	Italy, Japan.
ASEA-ATOM (Sweden)	2	1,382	2	BWR	Finland.
of which in service	(1)	(691)	(4)		
TOTAL	121	68,269	100		
TOTAL in service	47	16,702	(100)		

Table 82. **FREE WORLD - REACTOR DISTRIBUTION AND INSTALLED CAPACITIES 1980-2000***

(giga watts)

	1980		1985	1990	1995	2000	
	GW	% of Total	GW	GW	GW	GW	% of Total
LWR	126.0	87.0	228.0	387.7	544.7	716.8	86.1
HWR	7.6	5.2	14.9	25.9	43.6	73.5	8.8
AGR	2.9	2.0	4.9	8.0	12.4	16.8	2.0
GG	7.1	4.9	6.7	4.2	0.8	0.0	-
HTR	0.3	0.2	0.6	1.6	2.6	3.6	0.4
FBR	0.5	0.3	2.0	5.4	12.7	21.8	2.6
TOTAL	144.4	100.0	237.1	432.8	616.8	832.5	100.0

Source: INFCE - International Fuel Cycle Evaluation

LWR = Light Water Reactor
HWR = Heavy Water Reactor
AGR = Advanced Gas Cooled Reactor
GG = Graphite Gas Reactor
HTR = High Temperature Gas Cooled Reactor
FBR = Fast Breeder Reactor

* The table shows the lowest point of the range produced by the INFCE study group, but even this seems optimistic.

Table 83. NUCLEAR PLANT PERFORMANCE 1975-78

		Load Factor %	
		Mean	Range
Pressurised Water Reactor: (PWR)	1975	66	0-97
	1976	61	0-96
	1977	67	0-96
	1978	68	0-95
Boiling Water Reactor: (BWR)	1975	52	0-87
	1976	59	14-87
	1977	51	0-87
	1978	60	0-88
Pressurised Heavy Water Reactor: (PHWR)	1975	61	24-86
	1976	80	44-93
	1977	74	26-95
	1978	75	9-97
Gas Cooled Reactor: (GCR)	1975	65	15-84
	1976	65	36-82
	1977	65	44-78
	1978	62	38-84

Source: Sussex University Science Policy Research Unit

Notes: Nuclear plant is normally run at full load all the time it is available. Because of this the load factor (obtained by allowing for all the different types of outage, i.e. when the reactor is not running for any reason) is a good measure of the reliability of the reactor.

PHWR, built by Canada - came out best

Table 84. **FUEL USED FOR ELECTRICITY GENERATION 1979–2000**

(Excluding Africa, Middle East, China, Eastern Europe and USSR)

(million tonnes oil equivalent)

	1979		1990		2000		
	MTOE	% of Total	MTOE	% of Total	MTOE	% of Total	% Change 1979–2000
Oil	242	18	223	11	144	5	– 40
Gas	149	11	201	10	228	8	+ 53
Coal	496	37	736	37	1,042	38	+110
Fossil Fuels	887	66	1,160	58	1,414	51	-
Hydro etc.	307	23	412	21	540	20	+ 15
Nuclear	139	10	418	21	785	29	+465
TOTAL	1,334	100	1,900	100	2,739	100	+105
Total fuel use	4,545.1	100	5,836.0	100	7,317	100	+ 61
Fuel for electricity generation	1,334	29.3	1,990	34.1	2,739	37.4	+105

Source: Own Calculations.

Table 85. **COAL FIRED PERCENTAGE OF ELECTRICAL CAPACITY 1977 AND 2000**

	1977	2000
Japan	4	16
Western Europe	36	36
North America	36	48
Australia	64	84
TOTAL OECD	32	39

Source: Own Calculations / Various Sources

Table 86. **WORLD DEPENDENCE ON NUCLEAR ELECTRICITY BY SELECTED COUNTRY**

(%)	1976	1977	1980
Belgium	21	22.4	-
Sweden	18	21.7	24.5
Switzerland	18	16.8	-
France	-	13.4	26.0
U.K.	13	14.0	-
West Germany	-	11.0	-
United States	-	12.0	-
Bulgaria	-	-	20.0

Source: Atomic Industrial Forum / U.K.: Department of Energy Estimates

Table 87. **CHANGES IN EUROPEAN ELECTRICITY CONSUMPTION 1978-79**

Country	% Increase 1978-9	Country	% Increase 1978-9
Austria	+ 3.5	Netherlands	+ 4.5
Belgium	+ 6.1	Norway	+ 8.2
Czechoslovakia	- 1.0	Poland	+ 2.5
Finland	+ 7.5	Portugal	+ 10.1
West Germany	+ 6.0	Rumania	+ 2.1
France	+ 6.6	Sweden	+ 5.6
East Germany	+ 1.5	Switzerland	+ 4.0
Greece	+ 5.2	Turkey	+ 6.0
Hungary	+ 1.6	USSR	+ 3.7
Italy	+ 4.9	Yugoslavia	+ 5.3

Source: UK Electricity Council

Table 88. **USA – ENERGY FACTS 1979**

(million tonnes oil equivalent)

	Reserves	Consumption	Production
Oil	3,572	762.9	423.7
Gas	5,064	498.8	478.0
Coal	75,040	384.1	425.1[1]
Hydro etc.	n.a.	80.1	
Nuclear	*[2]	72.2	
TOTAL	83,676	1,898.1	

Population 1979 = 219.03 million

Per capita consumption = 8.7 tonnes of oil per annum.

Source Own Calculations.

Notes: [1] Estimate

[2] 708,000 tonnes uranium.

Table 89. UNITED STATES – CONSUMPTION BY PRIMARY FUEL 1969–1979

(Unit: Million tonnes of oil equivalent)

	OIL		GAS		COAL		HYDRO-POWER		NUCLEAR		TOTAL	
	MTOE	% of total	MTOE	% of total	MTOE	% of total	MTOE	% of total	MTOE	% of total	MTOE	% of total
1969	667.8	41	561.2	34	327.1	20	66.7	4	3.7	-	1,626.5	100
1970	694.6	42	564.1	34	329.5	20	65.9	4	5.7	-	1,659.8	100
1971	719.3	42	584.2	34	316.3	19	70.9	4	9.9	1	1,700.6	100
1972	775.8	44	587.4	33	316.6	18	72.8	4	15.2	1	1,767.8	100
1973	818.0	45	572.3	31	335.0	18	75.6	4	21.8	1	1,822.7	100
1974	782.6	44	555.1	31	331.9	19	70.7	4	28.8	2	1,769.1	100
1975	765.9	44	508.7	29	322.9	19	80.2	5	44.4	3	1,722.1	100
1976	822.4	45	516.4	28	346.5	19	73.8	4	51.6	3	1,810.7	100
1977	865.9	47	502.3	27	356.3	19	58.9	3	67.8	4	1,851.2	100
1978	888.8	47	504.2	26	355.2	19	79.8	4	75.5	4	1,903.5	100
1979	862.9*	45	498.8	26	384.1	20	80.1	4	72.2	4	1,898.1	100
Yearly change 1969–79	+2.6		–1.2		+1.6		+1.8		+34.8		+1.6	
Yearly change	+2.0		–2.1		+3.0		+2.5		+20.2		+1.4	
1990	720	38	496	21	690	30	87	4	174	7	2,330	100
2000	542	31	481	16	1,152	39	114	4	286	10	2,949	100

*1979 Imports of Crude & Products = 419.8 (48.6 % of oil consumption).

Source: Own Calculations / BP

Table 90. UNITED STATES PRIMARY FUEL MARKET 1979–2000

(million tonnes of oil equivalent)

	1979		1990		2000		
	MTOE	% of Total	MTOE	% of Total	MTOE	% of Total	% Change 1974-2000
Oil (Total)	862.9	45.5	840	36.0	733	24.8	– 15.0
Oil (home produced)[1]	483.1	25.4	310	13.3	305	10.3	– 37.0
Oil (imports)	419.8	22.1	470	20.2	370	12.5	– 12.0
Oil and gas (imports)	438.1	23.0	525	22.5	428	14.5	– 2.3
Gas (imports)	18.5	1.0	55	2.4	58	2.0	+213.0
Gas (home produced)	480.3	25.3	441	18.9	423	14.3	– 11.9
Gas (total)	498.8	26.3	496	21.3	481	16.3	– 3.6
Synthetics	-	-	43	1.8	183	6.2	-
Coal (total)	384.1	20.2	690	29.6	1,152	39.1	+200.0
Coal (home produced)	425.1*	22.4	553	23.7	1,262	42.8	+196.9
Hydro and other	80.1	4.2	87	3.7	114	3.9	+ 80.0
Nuclear	72.2	3.6	174	7.5	286	9.7	+296.0
Total Consumption	1,898.1	100.0	2,330	100.0	2,949	100.0	+ 55.0

Source: Own Calculations / Various Sources

Notes: [1] Including natural gas liquids

* Estimate

Table 91. JAPAN – CONSUMPTION BY PRIMARY FUEL 1969–1979

(Unit: Million tonnes of oil equivalent)

	OIL		GAS		COAL		HYDRO-POWER		NUCLEAR		TOTAL	
	MTOE	% of total	MTOE	% of total	MTOE	% of total	MTOE	% of total	MTOE	% of total	MTOE	% of total
1969	169.0	67	2.4	1	60.5	24	20.9	8	0.3	-	253.1	100
1970	199.1	69	3.6	1	62.6	22	21.0	7	1.2	-	287.5	100
1971	219.7	72	3.8	1	56.1	18	21.9	7	2.1	1	303.6	100
1972	234.4	74	3.7	1	57.1	18	20.0	6	2.2	1	317.4	100
1973	269.1	76	5.3	1	60.9	17	17.3	5	2.3	1	354.9	100
1974	258.9	73	7.0	1	63.8	18	18.7	5	4.6	1	353.0	100
1975	244.0	72	8.5	2	62.1	18	20.0	6	5.3	2	339.9	100
1976	253.5	72	10.4	3	59.9	17	21.4	6	9.0	2	354.2	100
1977	260.4	73	13.0	4	57.3	16	18.2	5	6.9	2	355.8	100
1978	262.7	72	17.0	5	54.0	15	17.4	5	12.5	3	363.6	100
1979	265.4	73	22.1	6	58.6	16	19.9	5	14.7	4	380.7	100
% change p.a. 1969–77	+4.6		+25		–0.3		–0.5		+48.4		+4.2	
% change p.a. 1974–79	+0.5		+25.7		–1.7		+1.3		+26.5		+1.5	

Source: BP.

Table 92. **JAPAN – CONSUMPTION BY PRIMARY FUEL 1979–2000**

(million tonnes oil equivalent)

	1979		1990		2000		
	MTOE	% of Total	MTOE	% of Total	MTOE	% of Total	% Change 1979–2000
Gas & Coal imports	62.1	16	121	24	214	33	+ 165
Gas	22.1	6	52	10	124	19	+ 461
Total coal	58.6	15	69	14	90	14	+ 53
Oil imports	265.4	70	292	58	264	41	-
Other (hydro etc.)	19.9	5	45	9	57	9	+ 186
Nuclear	14.7	4	44	9	105	16	+ 546
TOTAL	380.7	100	502	100	640	100	+ 65

Source: Own Calculations.

Table 93. **JAPAN - SOURCES OF OIL IMPORTS**

(%)

	1979	1980 (Jan-June)
Saudi Arabia	26.9	28.0
Indonesia	14.5	14.3
Iran	13.0	11.6
UAE	10.2	12.5
Kuwait	7.8	3.1
Iraq	6.1	8.1
Neutral Zone	5.9	5.3
Malaysia / Brunei	5.8	6.0
Oman	3.5	3.1
China	3.1	3.1
Others	3.2	4.9
TOTAL	100.0	100.0

Source: Various Sources

Table 94. WESTERN EUROPE – PRIMARY ENERGY CONSUMPTION 1969–1979

(million tonnes oil equivalent)

	Oil	**Gas**	**Coal**	**Hydro**	**Nuclear**	**TOTAL**
1969	565.5	53.8	302.7	86.3	10.7	1,019.0
1970	627.0	72.6	292.4	89.3	11.3	1,092.6
1971	656.0	93.4	269.1	90.4	13.2	1,122.1
1972	701.8	113.5	246.9	92.0	15.9	1,170.1
1973	748.9	129.9	254.8	91.9	16.6	1,242.1
1974	699.3	147.3	249.2	96.4	18.6	1,210.8
1975	664.4	153.5	236.2	97.1	25.3	1,176.5
1976	710.3	163.6	249.3	91.7	29.2	1,244.1
1977	697.3	169.9	248.2	110.1	35.6	1,261.1
1978	715.7	174.6	248.7	104.3	36.1	1,279.4
1979	726.5	185.5	264.0	108.3	41.3	1,325.7
% change per annum 1969–79	+2.5	+13.2	+1.4	+2.3	+14.4	+2.7
% change per annum 1974–79	+0.8	+4.7	+1.1	+2.3	+17.3	+1.8

Source: BP

Table 95. WESTERN EUROPE – CONSUMPTION BY PRIMARY FUEL 1979–2000

	1979		1990		2000		
	MTOE	% of Total	MTOE	% of Total	MTOE	% of Total	% Change 1979–2000
Gas & coal imports	79.5	6	198	12	331	18	+ 316
Of which: Gas[1]	25.7	2	72	4	137	7	+ 433
Coal	53.8	4	126	8	194	10	+ 261
Oil imports	647.1	49	544	33	510	27	– 21
Total imports	721.6	54	742	45	841	45	+ 16
Oil (indigenous)	79.4[2]	6	175	11	142	8	+ 79
Total oil consumption	726.5	55	719	44	707	38	– 3
Gas (indigenous)	159.8	12	185	11	137	7	– 14
Total gas consumption	185.5	14	257	16	274	15	+ 48
Hydro and other	108.3	8	115	7	121	6	+ 12
Coal (indigenous)	184.5	14	246	15	273	15	+ 48
Total coal consumption	264.0	20	372	23	457	24	+ 77
Nuclear	41.3	3	181	11	314	17	+ 660
TOTAL	1,325.6	100	1,644	100	1,867	100	+ 41

Source: Own Calculations.

Notes: 1 American Gas Association Transmission Conference

2 Oil production in 1979 was 115.9 MT, exports were 25.8 MT.

Table 96. UK – PRIMARY ENERGY CONSUMPTION 1969–2000

	OIL		COAL		GAS		WATER		NUCLEAR		TOTAL	
	MTOE	% of total	MTOE	% of total	MTOE	% of total	MTOE	% of total	MTOE	% of total	MTOE	% of total
1969	97.3	46.7	96.3	46.2	5.9	2.8	1.2	0.6	7.5	3.6	208.2	100
1970	103.6	48.1	92.3	42.9	11.2	5.2	1.5	0.7	6.7	3.1	215.3	100
1971	104.3	48.8	82.9	38.8	18.1	8.5	1.1	0.5	7.1	3.3	213.5	100
1972	110.5	51.2	72.9	33.8	25.2	11.7	1.2	0.6	6.5	3.0	215.6	100
1973	113.2	50.3	78.4	34.9	26.1	11.6	1.2	0.5	5.9	2.6	224.8	100
1974	105.3	49.1	69.1	32.2	31.8	14.8	1.2	0.6	7.1	3.3	214.5	100
1975	92.0	45.0	71.9	35.2	32.9	16.1	1.2	0.6	6.3	3.1	204.3	100
1976	91.4	44.1	72.7	35.0	34.6	16.7	1.1	0.5	7.6	3.7	207.4	100
1977	92.0	43.4	73.4	34.6	36.9	17.4	1.2	0.6	8.4	4.0	211.9	100
1978	94.0	44.5	70.4	33.3	37.9	17.9	1.2	0.6	7.9	3.7	211.4	100
1979	94.1	42.6	76.1	34.5	41.2	18.7	1.3	0.6	8.1	3.7	220.8	100
1990	83	36.4	77	33.8	48	21.0	3.1	1.3	1.9	7.5	230.1	100
2000	76	30.2	93	36.9	42	16.7	3.1	1.2	3.8	15.1	252	100
% increase 1979–2000	–19.2		+22.2		+1.9		+13.8		+36.9		+14.1	

Source: BP / Own Calculations.

Table 97. UK PRIMARY ENERGY CONSUMPTION 1979–2000

	1979		1990		2000		Average % change per year		
	MTOE	% of total	MTOE	% of total	MTOE	% of total	1969-79	1974-79	1979-2000
Oil	94.1	42.6	83	36.4	76	30.2	– 0.3	–2.2	–1.0
Coal	76.1	34.5	77	33.8	93	36.9	– 2.3	+1.9	+1.0
Gas	41.2	18.7	48	21.0	42	16.7	+21.5	+5.3	-
Water	1.3	0.6	3.1	1.3	3.1	1.2	+ 1.2	+1.3	+4.5
Nuclear	8.1	3.7	19	7.5	38	15.1	+ 0.8	+2.6	+3.0
TOTAL	220.8	100	230.1	100	252	100	+ 0.6	+0.6	+0.6

Source: Own Calculations.

Table 98. UK ENERGY USE TRENDS 1978-79, 1979-80 BY FINAL CONSUMER

	1978-1979 % Change	1978 %	1979 %	1979-1980 (Jan-Sept) % Change
Iron and Steel Industry:				
Coal	+ 54.0	1.1	1.5	- 6.0
Other solid fuel[1]	+ 9.8	48.0	49.2	- 35.8
Other coal derived fuels[2]	- 7.2	6.8	5.9	- 30.4
Gas[3]	+ 20.6	9.6	10.8	- 15.1
Electricity	+ 2.7	9.6	9.1	- 24.3
Petroleum	+ 0.6	24.9	23.4	- 44.5
TOTAL	+ 7.2	100.0	100.0	- 34.4
Other Industries:				
Coal	+ 7.0	12.3	12.8	- 20.9
Other solid fuel[1]	+ 2.8	0.8	0.8	- 29.4
Other coal derived fuels[2]	+ 1.4	0.4	0.4	+ 6.7
Gas[3]	+ 2.0	31.3	31.2	- 8.4
Electricity	+ 4.5	13.6	13.9	- 7.7
Petroleum	+ 0.1	41.6	40.7	- 15.6
TOTAL	+ 2.2	100.0	100.0	- 12.8
Transport Sector:				
Coal and other solid fuel	- 4.8	0.1	0.1	- 66.7
Electricity	-	0.7	0.7	-
Petroleum	+ 2.3	99.1	99.1	- 1.0
TOTAL	+ 2.3	100.0	100.0	- 1.1
Domestic Sector:				
Coal	+ 2.8	19.3	18.5	- 23.8

Continued/............

Table 98. continued.........

	1978-1979 % Change	1978 %	1979 %	1979-1980 (Jan-Sept) % Change
Other solid fuel[1]	- 1.3	5.0	4.6	- 18.2
Gas[3]	+ 13.3	47.2	49.8	+ 4.9
Electricity	+ 4.5	19.0	18.5	+ 1.2
Petroleum	- 2.0	9.3	8.5	- 22.6
TOTAL	+ 7.4	100.0	100.0	- 7.7
Other Final Users[4]:				
Coal	+ 0.6	6.7	6.5	- 20.0
Other solid fuel[1]	- 3.3	2.0	1.8	- 14.7
Gas[3]	+ 12.8	22.7	24.7	+ 4.0
Electricity	+ 5.6	23.2	23.7	+ 2.9
Petroleum	- 0.1	45.3	43.5	- 18.2
TOTAL	+ 4.1	100.0	100.0	- 7.8

Source: Department of Energy 'Energy Trends'

Notes: 1 Coke and other manufactured solid fuels

2 Coke oven gas, creosote/pitch mixtures and other liquid fuels derived from coal

3 Plus town gas

4 Mainly public administration, commerce and agriculture

Table 99 U.K. ENERGY DEMAND BY SECTOR 1977-2000

	1977		1990		2000		
	MTOE	% of Total	MTOE	% of Total	MTOE	% of Total	% Change 1977-2000
Domestic	56.0	26.3	59.0	25.7	59	23.3	+ 5
Iron and steel	18.0	8.6	19.0	8.4	19	7.7	+ 5
Other industry	61.0	28.8	66.0	28.8	81	32.2	+ 33
Transport	48.0	22.6	56.0	24.3	60	23.7	+ 25
Other consumers	28.0	13.4	29.0	12.6	33	13.2	+ 18
TOTAL	211.9	100.0	230.1	100.0	252	100.0	+ 19

Source: Own Calculations / Department of Energy

Table 100. ELECTRICITY SALES[1] BY SECTOR 1974-79 (%)

	% Iron and Steel Industry	% Other Industries	% Domestic	% Other[2]	% Total	Index 1974 = 100
1974	5.2	32.5	42.4	20.0	100	100
1975	5.1	32.5	40.9	21.4	100	99.7
1976	5.7	33.9	38.5	21.8	100	101.0
1977	5.5	33.8	30.1	22.5	100	103.2
1978	5.8	33.7	37.3	23.2	100	105.3
1979	5.7	33.6	37.2	23.4	100	110.2
% change 1978-79	+2.9	+4.5	+4.5	+5.2	+4.6	
1979 (Jan-June)	5.6	32.2	39.0	23.2	100	
1980 (Jan-June)	3.9	33.9	37.8	24.3	100	103.4[3]
% change 1974–79	+22.0	+14.2	–3.2	+28.9	+10.2	

1. Gigawatt hours
2. Mainly commerce, public administration and agriculture.
3. Own estimate for 1980.

Source: Department of Energy / Own Estimate.

Table 101. U.K. FUEL USED FOR ELECTRICITY GENERATION 1976–2000

(%)

	Coal[2]	Oil[2,3]	Gas	Nuclear	Hydro	TOTAL	Index 1976 = 100
1976	70.4	15.4	2.3	10.3	1.5	100.0	100
1977	69.8	15.8	1.5	11.3	1.5	100.0	103
1978	70.2	15.7	0.5	10.7	1.6	100.0	104
1979	72.8	14.9	0.5	10.1	1.5	100.0	110
% change 1978-9	+9.8	- 7.0	- 36.6	+3.5	+6.8	+5.9	-
1979 (Jan-Oct)	73.0	14.8	0.5	10.1	1.4	100.0	-
1980 (Jan-Oct)	78.1	9.9	0.2	10.3	1.3	100.0	103*
% change 1979–80	+0.4	- 37.5	- 58.2	- 3.7	- 8.9	- 6.1	-
1990	69.0	7.0	-	22.2	2.0	100.0	118
2000	60.0	5.0	-	33.0	2.0	100.0	136

Source: Department of Energy.

Notes: [1] Including coke

[2] Including quantities used in the production of steam for sale

[3] Including oil used in gas turbine and diesel plant and for lighting up coal fired boilers

* Own estimate for 1980.

Table 102. U.K. OIL AND GAS RESERVES VERSUS PRODUCTION

(million tonnes of oil equivalent)

	Gas	Oil
Production to end 1979	279	184
Proven reserves	628	1,200
% of proven reserves produced	44	15

	Gas Estimated		Oil Estimated	
	Low	High	Low	High
Reserves remaining in present discoveries	628	1.260	1.200	2.400
Ultimately recoverable reserves	917	2.000	2.200*	4.400*
% of ultimately recoverable reserves produced	30	14	8	4
Production in 1979	33.9		77.9	
Years of production possible at present rate	27	59	28	56

Source: U.K. Department of Energy / Own Calculations

* Reserves originally present on the U.K. Continental Shelf

Table 103. UK PETROLEUM PRODUCT[2] CONSUMPTION 1976-1980 BY USAGE[1]

(%)

	Power Stations[3]	Gas Works	Iron and Steel Industry	Other Industries	Transport[4]	Domestic	Other[5]	TOTAL
1976	14.6	0.5	4.4	24.5	40.0	4.5	11.4	100.0
1977	14.9	0.4	4.0	24.1	40.2	4.5	11.8	100.0
1978	15.6	0.5	3.8	23.5	41.3	4.4	11.0	100.0
1979	14.9	0.5	3.8	23.5	42.1	4.3	11.0	100.0
% change 1978-79	- 3.9	+ 4.3	+ 0.9	+ 0.3	+ 2.3	- 2.1	- 0.1	+ 0.4
1979 (Jan-Sept)	14.6	0.5	3.8	23.6	42.1	4.3	11.1	100.0
1980 (Jan-Sept)	10.1	0.5	2.4	25.6	50.3	3.6	10.5	100.0
% change 1979-80	- 14.4	- 19.9	- 46.2	- 18.9	+ 1.0	- 28.8	- 19.8	- 15.4

Source: Department of Energy / Own Calculations

Notes: 1 Calendar months
2 Excludes non-energy products and non energy use of naphtha (LDF)
3 Public supply, railway and transport power stations
4 Including fishing, coastal and inland shipping
5 Mainly public administration, commerce and agriculture

Table 104. U.K. (CONTINENTAL SHELF AND ON SHORE) OIL AND GAS PRODUCTION COMPARED WITH U.K. TOTAL PRIMARY FUEL CONSUMPTION 1975-1979

(%)

	Oil Production	Natural Gas Production	Total Consumption
1975	0.8	15.6	100
1976	5.8	16.2	100
1977	25.5	16.0	100
1979	35.1	15.3	100

Source: Development of the oil and gas resources of the United Kingdom (Department of Energy)

Table 105. U.K. CONTINENTAL SHELF AND ON SHORE OIL PRODUCTION 1979

	Million Tonnes
Offshore crude[1]	76.5
Heavier natural gases[2]	0.8
Condensate[3]	0.4
Onshore crude	0.1
TOTAL	77.9

Source: Development of the Oil and Gas Resources of the U.K.

Notes: [1] Crude oil includes condensate and dissolved gases present in the disposals of stabilised crude by the industry

[2] Heavier natural gases are ethane, propane and butane produced in the treatment of liquid or gaseous hydrocarbons at pipeline terminals

[3] Condensate which is a mixture of pentane and higher hydrocarbons arises mainly from the treatment of gas produced from the Frigg and Southern Basin fields

Table 106. **U.K. OFFSHORE OIL PRODUCTION 1975–1979 BY FIELD**

(million tonnes)

	1975		1976		1977		1978		1979		Total to end 1979	
	MT	% of Total	MT	% of Total	MT	% of Total	MT	% of Total	MT	% of Total	MT	% of Total
Argyll	0.5	45	1.1	9	0.8	2	0.7	1	0.8	1	3.9	2
Auk	-	-	1.2	10	2.3	6	1.3	2	0.8	1	5.7	3
Beryl	-	-	0.4	3	3.0	8	2.6	5	4.7	6	10.7	6
Brent	-	-	0.1	1	1.3	3	3.8	7	8.8	11	14.0	8
Claymore	-	-	-	-	0.3	1	3.0	6	4.0	5	7.4	4
South Cormorant	-	-	-	-	-	-	-	-	0.1	-	0.1	-
Dublin	-	-	-	-	-	-	0.7	1	5.7	7	6.3	3
Forties	0.6	54	8.6	74	20.1	54	24.5	46	24.5	32	78.3	44
Heather	-	-	-	-	-	-	0.1	1	0.8	1	1.0	1
Montrose	-	-	0.1	1	0.8	2	1.2	2	1.3	2	3.4	2
Ninian	-	-	-	-	-	-	0.1	-	7.7	10	7.8	4
Piper	-	-	0.1	1	8.6	23	12.2	23	13.2	17	34.1	19
Statfjord U.K.	-	-	-	-	-	-	-	-	0.1	-	0.1	-
Thistle	-	-	-	-	-	-	2.6	5	3.9	5	6.5	4
TOTAL	1.1	100	11.6	100	37.3	100	52.8	100	76.5	100	179.3	100

Source: Development of the Oil and Gas Resources of the U.K.

Table 107. FORECAST OF U.K. CONTINENTAL SHELF AND ONSHORE OIL PRODUCTION 1980-1984

(million tonnes)

	Low	High
1980	80.5	
1981	80	95
1982	82	110
1983	85	115
1984	90	120

Source: Department of Energy

Table 108. **GAS PRODUCTION AND IMPORTS 1975-1980**

	Production	Imported*
1975	31.4	0.8
1976	33.0	0.9
1977	34.6	1.5
1978	32.9	4.4
1979	32.8	7.8
1980	30.8	9.3
% change 1979-80	- 6.1	+ 18.2

Source: Energy Trends

* Includes imports from the Norwegian sector of the Frigg gasfield

Table 109. COAL PRODUCTION TRADE AND STOCKS 1975-1980

(million tonnes oil equivalent)

	Production	Imports	Exports	Stocks
1975	77.2	3.0	1.3	18.7
1976	74.3	1.7	0.9	19.9
1977	73.3	1.5	1.2	18.9
1978	74.2	1.4	1.3	20.7
1979	73.4	2.6	1.4	16.7
1980	78.1	4.4	2.4	22.6
% change 1979-80	+ 6.3	+ 69.0	+ 72.8	+ 35.2

Source: Department of Energy

Table 110. **U.K. TRADE BALANCE 1974–1979 (CRUDE AND PRODUCTS)**

(£ million)

	Imports		Exports		Balance	
	Crude Oil	Total	Crude Oil	Total	Crude Oil	Total Oil products
1974	3,726	4,623	29	768	–3,697	–3,855
1975	3,371	4,306	30	814	–3,341	–3,492
1976	4,445	5,641	177	1,254	–4,268	–4,387
1977	3,971	5,226	915	2,080	–3,056	–3,146
1978	3,506	4,799	1,241	2,374	–2,265	–2,425
1979	3,671	5,769	2,713	4,318	– 958	–1,451
1979 1st quarter	797	1,272	438	765	– 359	507
2nd quarter	821	1,341	642	966	– 179	– 375
3rd quarter	926	1,423	794	1,231	– 132	– 192
4th quarter	1,127	1,733	839	1,356	– 288	– 377
1980 1st quarter	1,196	1,922	930	1,465	– 266	– 457
2nd quarter	1,084	1,749	972	1,525	– 112	– 224
3rd quarter	935	1,539	1,071	1,643	+ 136	+ 104

Source: Department of Energy / Own Calculations.

Table 111. WEST GERMANY – PRIMARY ENERGY CONSUMPTION 1969–2000

	OIL		COAL		GAS		WATER		HYDRO		TOTAL	
	MTOE	% of total	MTOE	% of total	MTOE	% of total	MTOE	% of total	MTOE	% of total	MTOE	% of total
1969	117.1	32.5	92.2	41.4	9.2	4.1	3.2	1.4	1.2	0.5	222.9	100
1970	128.6	54.3	89.6	37.8	13.0	5.5	4.0	1.7	1.5	0.6	236.7	100
1971	133.5	56.4	83.8	35.4	16.9	7.1	2.9	1.2	1.4	0.6	238.5	100
1972	140.9	56.9	80.2	32.3	21.5	8.7	2.9	1.2	2.2	0.9	247.7	100
1973	149.7	56.5	82.1	31.0	27.0	10.2	3.3	1.2	2.8	1.1	264.9	100
1974	134.3	52.5	82.5	32.2	32.5	12.7	3.8	1.5	2.9	1.1	256.0	100
1975	128.9	53.1	70.7	39.1	34.4	14.2	3.6	1.5	5.0	2.1	242.6	100
1976	138.9	53.5	75.9	29.2	36.3	14.0	2.9	1.1	5.6	2.2	259.6	100
1977	137.1	52.8	71.7	27.6	38.9	15.0	3.7	1.4	8.4	3.2	259.7	100
1978	142.7	52.8	72.9	27.0	41.7	15.4	4.6	1.7	7.9	2.9	270.0	100
1979	146.9	51.5	78.6	27.6	45.9	16.1	4.0	1.4	8.1	2.8	285.0	100
Yearly change 1969–79	+2.3		–1.6		+17.4		+2.4		+23.1		+2.5	
Yearly change 1974–79	+1.8		–1.0		+7.1		+0.9		+27.2		+2.2	
1990	142	42	94	28	63	18	4.6	1	36	11	340	100
2000	136	35	116	30	68	18	4.6	1	59	15	384	100
% increase 1979-2000	–7		+48		+48		+15		+628		+35	

Source: Own Calculations.

Table 112. FRANCE – ENERGY CONSUMPTION 1969–2000

	OIL		COAL		GAS		HYDRO		NUCLEAR		TOTAL	
	MTOE	% of total	MTOE	% of total	MTOE	% of total	MTOE	% of total	MTOE	% of total	MTOE	% of total
1969	83.0	56.7	39.8	27.2	8.5	5.8	13.9	9.5	1.2	0.8	146.4	100
1970	94.3	59.8	37.8	24.0	9.3	5.9	14.9	9.4	1.4	0.9	157.7	100
1971	102.8	63.1	33.8	20.8	11.1	6.8	12.8	7.9	2.3	1.4	162.8	100
1972	114.1	66.6	29.5	17.2	13.2	7.7	11.0	6.4	3.6	2.1	171.4	100
1973	127.3	68.4	29.5	15.8	15.7	6.4	10.6	5.7	3.0	1.6	186.1	100
1974	121.0	65.7	30.3	16.5	17.2	9.3	12.6	6.8	3.0	1.6	184.1	100
1975	110.0	64.5	26.5	15.5	17.0	9.9	13.3	7.8	3.9	2.3	171.1	100
1976	119.5	65.0	30.0	16.4	19.0	10.4	10.6	5.6	4.1	2.2	183.2	100
1977	114.6	61.4	29.8	16.0	20.4	10.9	16.7	8.9	5.1	2.7	186.6	100
1978	119.0	62.0	30.5	15.9	20.9	10.9	15.0	7.8	6.4	3.3	191.8	100
1979	118.1	60.4	29.9	15.3	23.3	11.9	14.5	7.4	9.6	4.9	195.4	100
Yearly change 1969–79	+3.6		–2.8		+10.6		+0.4		+23.4		+2.9	
Yearly change 1974–79	–0.5		–0.3		+6.3		+2.8		+26.2		+1.2	
1990	92	39.1	38	16.2	26	11.1	20	8.5	59	25.1	235	100
2000	66	25.0	49	18.6	38	14.4	23	8.7	89	33.8	263	100
% change 1979–2000	–44		+64		+63		53		+827		+35	

Source: Own Calculations.

Table 113. ITALY – ENERGY CONSUMPTION 1969–2000

(Unit: Million tonnes of oil equivalent)

	OIL		GAS		COAL		HYDRO		NUCLEAR		TOTAL	
	MTOE	% of total	MTOE	% of total	MTOE	% of total	MTOE	% of total	MTOE	% of total	MTOE	% of total
1969	77.3	69.3	11.2	10.0	9.8	8.8	11.8	10.6	0.4	0.3	110.5	100
1970	87.3	71.6	12.3	10.1	9.9	8.1	11.6	9.5	0.8	0.7	121.9	100
1971	93.8	73.3	12.5	9.8	9.5	7.4	11.2	8.8	0.9	0.7	127.9	100
1972	98.2	74.1	12.3	9.5	8.9	6.7	12.1	9.1	1.0	0.7	132.5	100
1973	103.6	75.1	14.4	10.4	9.0	6.5	10.1	7.3	0.8	0.6	137.9	100
1974	100.8	73.6	15.8	11.5	9.8	7.2	9.7	7.1	0.8	0.6	136.9	100
1975	94.5	71.1	18.0	13.5	9.8	7.4	9.7	7.3	0.9	0.7	132.9	100
1976	98.8	69.2	22.0	15.4	9.7	6.8	11.2	7.8	1.0	0.7	142.7	100
1977	96.1	67.3	21.6	15.1	9.6	6.7	14.5	10.2	0.9	0.6	142.7	100
1978	99.8	69.9	22.5	15.8	9.8	6.9	12.2	8.5	1.1	0.8	145.4	100
1979	101.2	68.5	22.9	15.5	10.0	6.8	12.4	8.4	1.3	0.9	147.8	100
Yearly change 1969–79	+2.7		+7.5		+0.2		+0.6		+11.6		+3.0	
Yearly change 1974–79	+0.1		+7.7		+0.4		+5.0		+9.5		+1.5	
1990	117	62	31	16	18	9	13	7	11	6	190	100
2000	121	52	41	10	27	12	16	7	26	11	231	100
% change 1979–2000	+3		+79		+70		+29		+1900		+56	

Source: Own Calculations.

Table 114. SPAIN – ENERGY CONSUMPTION 1969–2000

	OIL		GAS		COAL		HYDRO		NUCLEAR		TOTAL	
	MTOE	% of total	MTOE	% of total	MTOE	% of total	MTOE	% of total	MTOE	% of total	MTOE	% of total
1969	24.6	56.8	0.1	0.2	10.3	23.8	8.1	18.7	0.2	0.5	43.3	100
1970	28.1	60.4	0.1	0.2	10.8	23.2	7.3	15.7	0.2	0.4	46.5	100
1971	30.9	60.0	0.4	0.8	10.9	21.2	8.6	16.7	0.7	1.4	51.5	100
1972	32.5	59.1	1.0	1.8	11.1	20.2	9.2	16.7	1.2	2.2	55.0	100
1973	39.1	65.1	1.0	1.7	10.8	18.0	7.5	12.5	1.7	2.8	60.1	100
1974	41.1	64.4	1.3	2.0	11.3	17.7	8.2	12.8	1.9	3.0	63.8	100
1975	42.7	65.5	1.3	2.0	11.7	18.2	6.8	10.6	1.9	2.9	64.4	100
1976	48.3	69.3	1.5	2.1	12.2	17.5	5.8	8.3	1.9	2.7	69.7	100
1977	45.5	62.8	1.4	1.9	13.4	18.5	10.5	14.5	1.7	2.3	72.5	100
1978	47.0	62.6	1.5	2.0	13.9	18.5	10.7	14.2	2.0	2.7	75.1	100
1979	47.3	61.2	1.4	1.8	14.5	18.8	12.3	15.9	1.7	2.2	77.2	100
Yearly change 1969–79	+6.8		+32.7		+3.5		+4.3		+23.0		+6.0	
Yearly change 1974–79	+2.9		+2.4		+5.2		+8.5		–1.5		+4.0	
1990	46	42	2	2	28	26	13	12	19	18	108	100
2000	46	33	5	4	40	29	12	10	35	25	138	100
% change 1979–2000	–7		+257		+176		–2		+1959		+79	

Source: Own Calculations.

Table 115. **USSR ENERGY FACTS 1979**

(million tonnes of oil equivalent)

	Proved Reserves MTOE	Consumption MTOE	% of World Total	Production	Energy Balance +/- MTOE	Population millions	Per Capita Consumption tonnes/per capita
Oil	9,115	441	10	586	+145	-	-
Gas	26,200	307	44	349	+42	-	-
Coal*	59,000	342	22	361	+19	-	-
Water power	n.a.	45	n.a.	n.a.	n.a.	-	-
Nuclear	n.a.	13	n.a.	n.a.	n.a.	-	-
TOTAL	94,315	1,148	16	1,296	+206	270	4.25

Source: Own Calculations

*Economically recoverable hard coal.

Table 116. USSR – CONSUMPTION BY PRIMARY FUEL – 1969–1979

(Unit: Million tonnes of oil equivalent)

	OIL		GAS		COAL		HYDRO		NUCLEAR		TOTAL	
	MTOE	% of total	MTOE	% of total	MTOE	% of total	MTOE	% of total	MTOE	% of total	MTOE	% of total
1969	238.6	33	144.7	20	299.0	42	29.7	4	0.8	-	712.8	100
1970	263.0	35	159.0	21	293.9	39	32.1	4	0.9	-	748.9	100
1971	279.2	35	173.8	22	302.3	38	32.5	4	1.1	-	788.9	100
1972	302.9	38	182.8	23	317.0	40	31.7	4	2.1	-	836.5	100
1973	325.7	37	198.8	23	315.0	36	31.6	4	3.0	-	874.1	100
1974	358.5	39	210.8	23	316.2	34	34.1	4	4.5	-	924.1	100
1975	375.1	39	230.0	24	326.2	34	32.5	3	6.0	1	1,016.7	100
1977	399.6	38	271.2	26	341.2	32	37.9	4	10.7	1	1,060.6	100
1978	419.2	38	289.2	26	344.4	31	41.1	4	11.0	1	1,104.9	100
1979	441.0	38	307.0	27	342.5	30	45.0	4	12.5	1	1,148.0	100
Yearly change 1974–79	+4.2		+7.8		+1.6		+5.7		+22.7		+4.4	

Source: BP

Table 117. USSR CONSUMPTION BY PRIMARY FUEL 1979–2000

(million tonnes of oil equivalent)

	1979		1990		2000		% Change
	MTOE	% of Total	MTOE	% of Total	MTOE	% of Total	1979–2000
Oil	441.0	38	548	33	611	31	+ 38
Gas	307.0	27	505	31	530	27	+ 73
Coal	342.5	30	486	30	712	36	+108
Hydro	45.0	4	70	4	78	4	+ 73
Nuclear	12.5	1	29	2	47	2	+276
TOTAL	1,148.0	100	1,638	100	1,978	100	+ 72

Source: Own Calculations.

Table 118. EASTERN EUROPE* – ENERGY CONSUMPTION 1969–2000

	OIL		COAL		GAS		WATER		NUCLEAR		TOTAL	
	MTOE	% of total	MTOE	% of total	MTOE	% of total	MTOE	% of total	MTOE	% of total	MTOE	% of total
1969	48.8	15.8	227.4	73.6	30.1	9.7	2.4	0.8	0.1	-	308.8	100
1970	54.2	16.5	236.9	72.3	33.3	10.2	3.0	0.9	0.1	-	327.5	100
1971	61.3	18.7	241.2	73.6	38.5	11.2	3.4	1.0	0.2	0.1	344.6	100
1972	67.2	19.0	241.3	68.2	41.2	11.6	3.7	1.0	0.5	0.1	353.9	100
1973	75.1	21.0	235.0	65.8	42.1	11.8	4.0	1.1	1.0	0.3	357.2	100
1974	77.5	21.1	237.6	64.8	46.0	12.5	4.2	1.1	1.5	0.4	366.8	100
1975	83.3	21.5	246.9	63.7	51.3	13.2	4.8	1.2	1.4	0.4	387.7	100
1976	89.7	22.0	253.9	62.1	57.5	14.1	4.6	1.1	2.8	0.7	408.5	100
1977	96.2	22.7	260.7	61.5	58.4	13.8	5.4	1.3	2.9	0.7	423.6	100
1978	98.9	22.8	266.7	61.4	60.0	13.8	5.3	1.2	3.6	0.8	434.5	100
1979	101.1	22.9	270.0	61.1	61.5	13.9	5.4	1.2	3.8	0.9	441.8	100
Yearly change 1969–79	+7.6		+1.7		+7.4		+8.4		+42.4		+3.6	
early change 1974–79	+5.5		+2.6		+6.0		+5.2		+20.4		+3.8	
1990	114	19.9	358	62.5	84	14.7	7	1.2	9	1.6	572	
2000	117	16.5	463	65.5	101	14.3	9	1.3	17	2.4	707	

*excluding Yugoslavia.

Source: Own Calculations.

Table 119. OIL CONSUMPTION - EASTERN EUROPE AND THE COMMUNIST WORLD 1970-2000
(million tonnes of oil)

	1970		1975	1977	1978	1979		% Change 1975–79	1990		2000	
	M.T.	% of total	M.T.	M.T.	M.T.	M.T.	% of total		M.T.	% of total	M.T.	% of total
Bulgaria	8.5	15.7	13.1	13.6	14.4	14.4	14.2	+9.9	16	14.0	16	13.8
E. Germany	10.1	18.6	14.7	17.1	17.9	18.4	18.2	+25.2	21	18.4	21	18.4
Poland	8.8	16.2	14.5	17.6	16.9	16.5	16.3	+13.8	19	16.7	19	16.3
Rumania	10.3	19.0	13.9	17.6	18.9	18.9	18.7	+34.5	21	19.1	22	19.2
Czechoslovakia	9.8	18.1	16.0	19.4	19.9	20.3	20.1	+26.9	23	20.2	23	20.1
Hungary	6.7	12.4	10.1	10.8	10.9	12.6	12.5	+24.7	14	12.4	15	12.7
TOTAL E. EUROPE	54.2	100	83.3	96.2	98.9	101.1	100	+21	114	100	117	
E. Europe	54.2	15.7	83.3	96.2	98.9	101.1	16.0	+21	114	14.9	117	13.8
USSR	263	76.1	375	399.6	419	441	69.7	+18	548	86.5	611	71.9
China	28.2	8.2	68.3	82.0	84.7	91.1	14.4	+33	101	15.9	122	14.3
COMMUNIST WORLD TOTAL	345.4	100	526.8	577.8	602.8	633.2	100	+20	763	100	850	100

*excluding Yugoslavia

Source: Own Calculations./ Le Petrole en Chiffres

Table 120. CHINA – ENERGY FACTS 1979

(million tonnes of oil equivalent)

	Reserves MTOE	Consumption	Potential Life of Indigenous Reserves at 1979 Consumption	Production MTOE	Energy Balance +/- MTOE	Population (1979 Estimate) millions	Per Capita Consumption Tonnes of oil Equivalent per capita
Oil	2,740	91.1	30.1	106.1	+ 15.0	-	-
Gas	645	66.8	9.6	†	†	-	-
Coal	56,280	410.0	137.3	†	†	-	-
Water power	n.a.	9.0	n.a.	9.0	n.a.	-	-
Nuclear	n.a.	-	n.a.	-	n.a.	-	-
TOTAL	59,665	576.9	103.0	n.a.	n.a.	941.27	0.61

Source: Own Calculations.

Table 121. CHINA – CONSUMPTION BY PRIMARY FUEL 1969–1979

(Unit: Million tonnes oil equivalent)

	OIL		GAS		COAL		HYDRO		NUCLEAR		TOTAL	
	MTOE	% of total	MTOE	% of total	MTOE	% of total	MTOE	% of total	MTOE	% of total	MTOE	% of total
1969	20.4	8	1.9	1	227.4	89	6.0	2	-	-	235.7	100
1970	28.2	10	3.3	1	252.0	87	6.2	2	-	-	289.7	100
1971	36.7	11	5.2	2	273.2	85	6.4	2	-	-	321.5	100
1972	43.1	13	8.3	2	278.1	83	6.6	2	-	-	336.1	100
1973	53.8	15	12.5	3	292.5	80	6.8	2	-	-	365.6	100
1974	61.9	16	17.5	4	307.4	78	7.1	2	-	-	393.9	100
1975	68.3	16	22.7	5	321.8	77	7.4	2	-	-	420.2	100
1976	76.9	17	27.3	6	337.5	75	7.7	2	-	-	449.4	100
1977	92.0	17	30.0	6	354.4	75	8.0	2	-	-	474.4	100
1978	84.7	17	33.0	6	380.0	75	8.0	2	-	-	505.7	100
1979	91.1	16	66.8	12	410.0	71	9.0	2	-	-	576.9	100
Average % change 1974–79	+8.00		+30.8		+5.9		+4.9		-		+7.9	

Source: BP.

Table 122. **ENERGY DEMAND GROWTH PREDICTION CHINA**

(% per year)

	1980-1985	1985-1990	1990-2000
Oil	0.9	1.1	1.9
Gas	5.0	4.1	0.1
Coal	5.9	3.6	3.3
Hydro	4.9	2.1	1.3
Nuclear	-	-	-

* Predicted population growth with implemention of present birth control legislation to average 0.9% per annum to year 2000.

* Population predicted to reach 1.136.1 million by 2000.

Source: Own Calculations

Table 123. CHINA CONSUMPTION BY PRIMARY FUEL 1979-2000

(million tonnes of oil equivalent)

	1979		1990		2000		% Change
	MTOE	% of Total	MTOE	% of Total	MTOE	% of Total	1979-2000
Oil	91.1	16	101	11	122.0	10	+ 34
Gas	66.8	12	113	12	114.0	9	+ 66
Coal	410.0	71	689	75	951.0	79	+ 132
Hydro	9.0	2	12	1	13.6	1	+ 51
Nuclear	-	-	-	-	-	-	-
TOTAL	576.9	100	915	100	1.201.0	100	+ 108

Source: Own Calculations

Table 124. **CANADA ENERGY FACTS 1979**

(million tonnes of oil equivalent)

	Proved Reserves	**Consumption**	**Production**	**Energy Balance +/-**	**Fossil Fuel Consumption Life (years)**
Oil	915	89.9	86.0	– 3.9	10
Gas	2,280	49.2	63.2	+14.0	46
Coal	2,842*[2]	21.6	25.0*[1]	+ 3.4*	132
Hydro	n.a.	54.2	54.2	n.a.	n.a.
Nuclear	n.a.	8.5	8.5	n.a.	n.a.
TOTAL Fossil fuel	6,037	223.4	236.9	+13.5	27

Source: Own Calculations.

[1] Estimate
[2] Mainly lignite

Table 125. CANADA – CONSUMPTION BY PRIMARY FUEL 1969–1979

(Unit: Million tonnes oil equivalent)

	OIL		GAS		COAL		HYDRO		NUCLEAR		TOTAL	
	MTOE	% of total	MTOE	% of total	MTOE	% of total	MTOE	% of total	MTOE	% of total	MTOE	% of total
1969	69.1	45	29.6	19	15.8	10	38.2	25	-	-	152.7	100
1970	73.0	45	32.9	20	16.9	10	40.1	25	-	-	162.7	100
1971	75.8	45	34.9	21	16.1	10	40.5	24	1.0	-	168.3	100
1972	79.3	44	39.3	22	15.2	8	44.2	25	1.7	1	179.7	100
1973	83.7	44	41.8	22	15.6	8	46.1	24	3.7	2	190.9	100
1974	84.8	43	42.2	21	15.9	8	51.1	26	3.6	2	197.6	100
1975	83.1	43	43.1	22	15.5	8	50.3	26	3.0	1	195.0	100
1976	85.9	42	46.1	22	18.3	9	52.3	25	4.1	2	206.7	100
1977	85.6	40	45.9	22	23.4	11	51.2	24	6.9	3	213.0	100
1978	86.9	41	47.3	22	19.2	9	56.0	26	8.5	4	217.9	100
1979	89.9	40	49.2	22	21.6	10	54.2	24	8.5	4	223.4	100
Yearly Change 1974–79	+1.2		+3.1		+6.3		+1.2		+18.8		+2.5	

Source: BP.

Table 126. CANADA CONSUMPTION BY PRIMARY FUEL 1979-2000

(million tonnes of oil equivalent)

	1979		1990		2000		% Change
	MTOE	% of Total	MTOE	% of Total	MTOE	% of Total	1979-2000
Oil	89.9	40	88	30	67	20	- 25
Gas	49.2	22	66	23	78	23	+ 58
Coal	21.6	10	36	12	55	16	+ 155
Hydro	54.2	24	62	21	70	21	+ 29
Nuclear	8.5	4	21	7	34	10	+ 300
Synthetics	-	-	24	8	53	16	n.a.
TOTAL	223.4	100	297	100	334	100	+ 49

Source: Own Calculations

Table 127. MIDDLE EAST PRIMARY ENERGY CONSUMPTION 1969–2000

(Unit: Million tonnes oil equivalent)

	OIL		COAL		GAS		WATER		NUCLEAR		TOTAL	
	MTOE	% of total		% of total	MTOE	% of total	MTOE	% of total	MTOE	% of total	MTOE	% of total
1969	47.2	80	-	-	11.3	19	0.6	1	-	-	59.1	100
1970	49.5		-		19.0		0.7		-		69.2	100
1971	54.0		-		19.0		0.9		-		73.9	100
1972	56.9		-		21.0		0.8		-		78.7	100
1973	62.2		-		24.1		0.8		-		87.1	100
1974	67.1		-		27.6		1.6		-		96.3	100
1975	66.8		-		26.2		1.6		-		94.6	100
1976	74.7		-		26.9		0.9		-		102.5	100
1977	78.9		-		28.8		1.0		-		108.7	100
1978	83.3	73	-	-	30.1	26	1.0	1	-	-	114.4	100
1979	74.8	71	-	-	30.0	28	1.0	1	-	-	105.8	100
Yearly change 1969–1979	+4.7				+10.2		+5.7				+6.0	
Yearly change 1974–1979	+2.2				+1.7		−7.6				+1.9	
1990	137	63	-	-	79	36	1.2	1	0.8	-	218	100
2000	350	72	-	-	134	27	1.4	-	1.4	-	487	100

Source: Own Calculations.

Table 128(a) MIDDLE EAST GAS CONSUMPTION 1979–2000 (million tonnes oil equivalent)

	1979	%	1990	%	2000	%
Iran	9.6	32	28	35	50.0	37
Saudi Arabia	6.8	23	21	27	36.0	27
Iraq	1.9	6	5	6	9.3	7
Kuwait	1.7	6	4	5	9.0	7
Others	10.0	35	21	27	29.0	22
TOTAL	30.0	100	79	100	134.0	100

Source: Own Calculations / Various Sources

Table 128(b) MIDDLE EAST OIL CONSUMPTION 1975-2000

	1975		1979		1980		1990		2000	
	MT	% of Total	MT	% of Total	MT	% of Total	MT	% of Total	MT	% of Total
Iran	19.1	31.0	n.a.	-	30.0	40.0	34.0	25.0	110.0	31.4
Saudi Arabia	6.8	11.0	n.a.	-	12.4	17.0	41.8	30.0	92.6	26.4
Iraq	7.0	11.0	n.a.	-	10.9	15.0	22.4	16.0	62.1	17.7
Kuwait	1.2	1.9	n.a.	-	1.9	2.5	4.7	3.4	11.7	3.3
UAE	1.2	1.9	n.a.	-	2.1	2.8	-	-	14.2	4.0
Others	27.0	43.0	n.a.	-	29.0	24.0	34.0	25.0	29.8	17.0
TOTAL	62.2	100.0	74.8	-	74.4	100.0	136.9	100.0	350.4	100.0

Source: Own Calculations / Various Sources

Table 129. **ESTIMATED OPEC PRODUCTION CAPACITY**

	MT Oil	% of Total	Proven Reserves	Life at Stated Production
Saudi Arabia	600	34	22,261	37.1
Iran	200	11	7,870	39.3
Iraq	200	11	4,159	20.8
Kuwait	150	9	9,007	60.0
Venezuela	125	7	2,550	20.4
Nigeria	100	6	2,348	23.4
Libya	100	6	3,086	30.9
UAE	100	6	3,865	38.6
Indonesia	75	4	1,306	17.4
Algeria	50	3	1,102	22.0
Qatar				
Gabon	50	3	711	14.2
Ecuador				
TOTAL	1,750	100	58,410	33.4

Estimated recoverable oil in the Middle East = 84 BTO

Life of reserves at stated production - 48.

Source: Own Calculations./ Various Sources

Table 130. LESS DEVELOPED COUNTRIES* OIL AND ENERGY DEMAND 1979–2000 AS A % OF WORLD DEMAND

Regions	1979		1990		2000	
	Oil	Total Energy	Oil	Total Energy	Oil	Total Energy
LDCs MTOE	504	881	761	1,495	1,214	2,447
%	16.2	12.7	21.3	15.7	29.5	19.9
World MTOE	3,119.6	6,960.4	3,571	9,506	4,109	12,296
%	100	100	100	100	100	100

*includes S. Africa

Source: Own Calculations./ Various Sources

Table 131. OPEC OIL PRODUCTION 1969-79

(Unit: million tonnes)

	S. AMERICA		MIDDLE EAST									AFRICA				S.E. ASIA
												North		West		
	Ecaudor	Venezuela	Iran	Iraq	Kuwait	Neutral Zone	Saudi Arabia	Qatar	Abu Dhabi	Dubai	Sharjah	Algeria	Libya	Gabon	Nigeria	Indonesia
1969	0.2	188.7	168.1	74.9	131.0	21.7	150.2	17.0	28.9	0.5	-	44.5	149.9	5.0	26.4	37.1
1970	0.2	195.2	191.3	76.9	139.1	26.0	178.0	17.7	33.4	4.3	-	48.5	159.8	5.4	52.9	42.2
1971	0.2	187.7	227.0	83.5	148.8	28.3	225.0	20.5	44.9	6.2	-	36.5	133.1	5.8	74.7	44.1
1972	3.8	171.5	251.9	72.1	153.0	29.3	287.2	23.2	50.6	7.6	-	49.8	108.2	6.3	88.9	53.4
1973	10.2	174.0	293.2	99.0	140.4	27.6	367.9	27.3	62.6	10.8	-	51.2	104.9	7.5	100.1	66.0
1974	8.7	158.5	301.2	96.7	116.3	28.0	412.4	24.9	67.7	12.0	1.4	47.1	73.3	10.0	112.2	67.9
1975	7.9	125.3	267.7	111.0	94.0	25.0	343.9	21.0	67.3	12.6	1.9	47.5	71.3	11.2	88.8	64.6
1976	9.1	122.9	295.0	118.8	98.2	24.4	421.6	23.4	76.8	15.6	1.8	50.1	93.3	11.2	102.9	74.6
1977	9.1	119.5	283.5	122.3	91.5	18.5	455.0	21.1	80.0	15.8	1.4	53.5	99.4	11.1	104.1	83.5
1978	10.0	115.4	260.4	127.6	97.0	23.9	409.8	23.4	69.7	18.0	1.1	57.2	95.2	10.8	95.1	81.0
1979	10.5	125.4	155.6	169.3	114.1	29.0	468.3	24.6	70.2	17.6	0.7	56.4	99.6	10.2	114.2	78.8
% of Total	0.7	8.1	10.1	11.0	7.4	1.9	30.3	1.6	4.5	1.1	0.04	3.6	6.4	0.7	7.4	5.1
Yearly Change 1969–79	+48.9	–4.0	–0.8	+8.5	–1.4	+2.9	+12.0	+3.7	+9.3	+42.8	-	+2.4	–4.0	+7.4	+15.8	+7.8
Yearly Change 1974–79	+3.9	–4.6	–12.4	+11.9	–0.4	+0.7	+2.6	–0.2	+0.7	+7.9	–12.9	+3.7	+6.3	+0.5	+0.4	+3.0
RESERVES	145	2,550	7,870	4,159	9,007		22,261	497	3,673	192	†	1,102	3,086	69	2,348	1,306

Source: BP / Le petrole en chiffres / Own Calculations.

Table 132. OPEC – OIL AND ENERGY CONSUMPTION – 1975–2000

(Unit: Million tonnes oil equivalent)

	PETROLEUM PRODUCTS						OTHER ENERGY USE IN 2000		
	1975	1980	1985	1990	1995	2000	Gas	Other	Total Energy
Algeria	3.9	8.2	13.5	22.2	36.6	60.2	16.7	2.4	79.3
Ecaudor	1.9	2.9	4.2	6.1	8.9	12.8	0.3	0.2	13.4
Gabon	0.6	1.0	1.6	2.6	4.4	7.0	-	-	7 0
Indonesia	12.6	19.2	29.6	45.6	70.1	108.0	11.0	9.0	127.9
Iran	19.1	30.0	30.0	34.	85.	110.0	50.	6.6	166.6
Iraq	7.0	10.9	14.1	22.4	38.3	62.1	9.3	-	71.4
Kuwait	1.2	1.9	3.0	4.7	7.4	11.7	9.0	-	20.7
Libya	2.4	4.5	8.4	15.6	28.8	53.3	13.9	-	67.2
Nigeria	5.3	8.6	13.3	20.7	32.1	49.9	17.3	4.2	71.5
Qatar	0.2	0.4	0.7	1.1	1.8	2.9	-	-	2.9
S. Arabia	6.8	12.4	22.8	41.8	76.4	92.6	36.0	-	175.8
U.A.E.	1.2	2.1	3.3	5.4	8.9	14.2	-	-	-
Venezuela	10.5	16.0	25.2	39.5	62.0	97.2	14.1	7.1	118.6
TOTAL OPEC	72.9	118.1	169.7	261.7	460.7	681.9	177.6	25.3	884.8

Source: Own Calculations./BP / Le Petrole en Chiffres / Various Sources

Table 133. AUSTRALIA AND NEW ZEALAND ENERGY FACTS

Australia population 14.71 million
New Zealand population 3.12 million

PRIMARY FUEL FACTS 1979

(million tonnes oil equivalent

	Oil	Coal	Gas	TOTAL
Reserves:				
Australia	300.0[1]	16,080	728.0)	17,271.0
New Zealand		†	163.0)	
Production:				
Australia	21.0	51[2]	6.6)	71.92
New Zealand	†	†	1.32)	
Imports:				
Australia	18.6[3]	-	-	18.6
New Zealand	†	†	†	†
Exports:				
Australia	1.1[3])	3)	)	23.1
New Zealand	)	22)	-)	
Consumption:				
Australia	38.0[3])	29[3])	8.4[3])	75.4
New Zealand	)	)	)	

Source: Own Calculations. † minimal

Notes: [1] Australia and New Zealand
[2] 1977
[3] Australasia

1969–2000

(Unit: Million tonnes oil equivalent)

	OIL		COAL		GAS		WATER		NUCLEAR		TOTAL	
		% of total		% of total		% of total		% of total		% of total		% of total
1969	27.9	51.0	21.2	38.8	0.2	-	5.4	9.9	-	-	54.7	100
1970	29.7	50.6	21.5	36.6	1.5	2.5	6.0	10.0	-	-	58.7	100
1971	31.3	50.6	21.3	34.4	2.2	3.5	7.1	11.5	-	-	61.9	100
1972	31.7	49.6	22.1	34.6	3.2	5.0	6.9	10.8	-	-	63.9	100
1973	34.8	52.2	22.9	34.4	3.9	5.9	5.0	7.5	-	-	66.6	100
1974	35.8	51.2	24.4	34.9	4.6	6.6	5.1	7.3	-	-	69.9	100
1975	35.1	44.1	26.1	36.9	4.9	6.8	5.4	7.5	-	-	71.5	100
1976	36.5	48.0	24.9	32.8	6.1	8.0	8.5	8.1	-	-	76.0	100
1977	38.0	47.3	26.8	33.4	7.4	9.2	8.1	10.1	-	-	80.3	100
1978	37.6	46.2	28.2	34.3	7.8	9.5	8.5	10.3	-	-	82.1	100
1979	38.0	44.8	29.0	34.2	8.4	9.9	9.4	11.1	-	-	84.8	100
Yearly change 1969-79	+3.1		+3.2		+42.2		+5.6		-	-	+4.8	-
Yearly change 1974-79	+1.2		+3.5		+12.8		+12.9		-	-	+3.9	-
1990	41	36	42	37	19	17	10.6	9	-	-	113	100
2000	38	26	60	40	37	25	13.0	9	-	-	148	100

*Australia, New Zealand, Papua New Guinea, South West Pacific Island.
Source: BP

Table 135. LATIN AMERICA ENERGY FACTS 1979
(Mexico, Caribbean (Including Puerto Rico), Central and South America)

Population (1980 Estimates): Mexico 70.02 million, Brazil 119.54 million, Venezuela 13.62 million, Argentina 26.83 million, Colombia 26.52 million, Ecuador 8.15 million, Peru 17.42 million.

PRIMARY FUEL FACTS 1979

(million tonnes oil equivalent)

	Oil	**Coal**	**Gas**
Reserves:			
Mexico	4,400	n.a.	1,490
Venezuela	2,550	n.a.	1,023
Argnetina	334	n.a.	396
Brazil	167	n.a.	39
Ecuador	145	n.a.	36
Colombia	101	n.a.	103
Trinidad	100	n.a.	172
Peru	87	n.a.	26
Chile	51	n.a.	60
Bolivia	49	n.a.	131
TOTAL	7,956	2,680	3,476
Consumption	211.8	16.0	44.0

Source: Various Sources

Table 136. LATIN AMERICA – PRIMARY ENERGY CONSUMPTION 1969–2000

(Unit: Million tonnes oil equivalent)

	OIL		COAL		GAS		WATER		NUCLEAR		TOTAL	
	MTOE	% of total	MTOE	% of total	MTOE	% of total	MTOE	% of total	MTOE	% of total	MTOE	% of total
1969	126.8	67	9.5	5	34.1	18	18.3	10	-	-	188.7	100
1970	137.2	69	10.2	5	30.4	15	21.1	11	-	-	198.9	100
1971	147.6	69	10.5	5	32.6	15	22.2	10	-	-	212.9	100
1972	151.5	68	11.3	5	36.4	16	23.9	11	-	-	223.1	100
1973	163.7	68	11.7	5	36.5	15	28.1	12	-	-	240.0	100
1974	171.3	67	13.1	5	37.8	15	32.0	13	0.2	-	254.4	100
1975	174.0	66	14.2	5	39.2	15	33.6	13	0.7	-	261.7	100
1976	185.7	67	15.2	5	40.5	15	35.9	13	0.7	-	278.0	100
1977	193.4	66	14.1	5	39.6	14	44.0	15	0.4	-	291.5	100
1978	201.4	66	15.2	5	42.3	14	43.6	14	0.7	-	306.2	100
1979211.8	211.8	66	16.0	5	44.0	14	46.8	15	0.8	-	319.4	100
Yearly change 1969–79 (%)	+5.3		+5.4		+2.6		+9.8		-		+5.4	
Yearly change 1974–79 (%)	+4.3		+4.1		+3.1		+7.9		+30.3		+4.6	
1990	284	58	36	7	67	14	86	18	13	3	486	100
2000	346*	51	53	8	102	15	150	22	23	3	674	100
% increase 1979-2000	+67		+231		+132		+220		+2,775		+111	

*includes synthetics.

Source: Own Calculations/ BP

Table 137. SOUTH ASIA ENERGY FACTS 1979
(million tonnes oil equivalent)

	Oil		Gas	Coal
Reserves:				
India	349		224	8,040
Afghanistan	n.a.		60	n.a.
Bangladesh	n.a.		228	n.a.
Pakistan	n.a.		384	n.a.
	Production	**Consumption**	**Production**	**Production**
India	13.0	22.0	1.6	48*
Afghanistan	n.a.	n.a.	2.1	n.a.
Bangladesh	n.a.	n.a.	n.a.	n.a.
Pakistan	n.a.	3.9	5.3	n.a.
TOTAL	14.9	36.7	n.a.	n.a.

Source: Various Sources

* 1977

Table 138. SOUTH ASIA – ENERGY CONSUMPTION 1969–2000

[Afghanistan, Bangladesh, Burma, India, Pakistan, Sri Lanka]

(Unit: Million tonnes oil equivalent)

	OIL	COAL	GAS	WATER	NUCLEAR	TOTAL
	MTOE	MTOE	MTOE	MTOE	MTOE	MTOE
1969	26.5	54.5	4.8	6.9	0.4	93.1
1970	26.8	51.7	5.4	7.4	0.	91.9
1971	27.9	52.4	6.0	8.1	0.5	94.9
1972	24.9	54.6	6.8	8.1	0.8	99.2
1973	31.3	55.0	8.0	8.0	1.0	103.3
1974	29.6	56.8	7.9	8.9	0.8	104.0
1975	30.1	59.1	8.1	7.4	1.0	105.7
1976	32.6	62.5	8.8	9.6	1.0	114.5
1977	34.5	64.0	9.4	10.2	0.8	118.9
1978	37.1	67.2	9.7	10.7	0.6	125.3
1979	36.7	74.9	6.4	10.7	0.6	129.3
Yearly change 1969–79 (%)	+3.3	+3.2	+3.0	+4.4	+6.3	+3.4
Yearly change 1974–79 (%)	+4.4	+5.7	–4.1	+3.7	–4.2	+4.5
1990	46	118	7.1	14.5	10	
2000	59	180	11.1	17.3	18	

Source: Own Calculations / BP

Table 139. **SE ASIA ENERGY FACTS 1979**

(Brunei, Kampuchea, Malaysia, Indonesia, Hong Kong, Korea, Lao PDR, Philippines, Singapore, Thailand, Taiwan, Vietnam)

(million tonnes oil equivalent)

	Oil	Coal	Gas
Reserves:			
Indonesia	1,306	n.a.	946
Malaysia	363	n.a.	413
Brunei	245	n.a.	n.a.
Taiwan	n.a.	n.a.	22
Thailand	n.a.	n.a.	185

	Oil		Coal		Gas	
	Production	Consumption	Production	Consumption	Production	Consumption
Indonesia	78.8	20.0	3*	n.a.	24.20	n.a.
Malaysia	21.9	5.0	n.a.	n.a.	0.09	n.a.
Brunei	11.9	n.a.	n.a.	n.a.	7.50	n.a.
Taiwan	n.a.	n.a.	n.a.	n.a.	2.00	n.a.
Philippines	n.a.	10.3	n.a.	n.a.	n.a.	n.a.
Singapore	n.a.	11.4	n.a.	n.a.	n.a.	n.a.
TOTAL	n.a.	183.5	n.a.	47.8	n.a.	7.1

Source: Various Sources * Estimate

Table 140. SOUTH EAST ASIA ENERGY CONSUMPTION 1969–2000

(million tonnes oil equivalent)

	Oil		Coal		Gas		Water		Nuclear		TOTAL	
	MTOE	% of Total	MTOE	% of Total	MTOE	% of Total	MTOE	% of Total	MTOE	% of Total	MTOE	% of Total
1969	53.4	59	29.3	32	2.0	2	5.6	6	-	-	90.3	100
1970	59.6		31.9		2.3		6.1		-	-	99.9	100
1971	63.9		33.8		2.3		6.3		-	-	106.3	100
1972	71.2		35.5		2.4		6.2		-	-	115.3	100
1973	77.6		36.4		3.6		6.4		-	-	124.0	100
1974	79.2		38.0		4.3		7.4		-	-	128.9	100
1975	81.2	61	40.4	30	3.1	3	7.3	5	-	-	133.0	100
1976	88.7	63	39.8	28	4.2	3	7.7	5	-	-	140.4	100
1977	95.8	63	43.1	28	4.6	3	7.7	5	-	-	151.2	100
1978	105.5	62	47.2	28	7.9	5	9.1	5	1.3	1	171.0	100
1979	116.9	64	47.8	26	7.9	4	9.5	5	1.4	1	183.5	100
Yearly change 1969-79	+8.2		+5.0		+14.7		+5.4		-	-	+7.3	100
Yearly change 1974-79	+8.1		+4.7		+12.9		+5.2		-	-	+7.3	100
1990	170.0	59	66.0	23	19.0	7	13.0	4	21.0	7	289.0	100
2000	199.0	50	105.0	27	39.0	-	16.0	10	36.0	9	395.0	100

Source: Own Calculations.

Table 141. AFRICA – ENERGY FACTS 1979

(million tonnes oil equivalent)

	Oil	**Coal**	**Gas**	**Hydro**
Reserves:				
Libya	3,086.0	†	598.0	n.a.
Nigeria	2,348.0	†	1,251.0	n.a.
Algeria	1,102.0	†	2,322.0	n.a.
Egypt	428.0	†	73.1	n.a.
Angola	167.0	†	34.4	n.a.
Gabon	69.0	†	21.5	n.a.
South Africa	384.0	13,400.0	-	n.a.
TOTAL	7,584.0	n.a.	4,418.0	n.a.
Production:				
Libya	99.6	†	3.9	†
Nigeria	114.2	†	1.72	†
Algeria	56.4	†	18.4	†
Egypt	25.5	†	1.0	†
Angola	9.0	†	0.1	†
Gabon	†	†	0.1	†
South Africa	†	40.9*	-	†
TOTAL	324.4	n.a.	25.5	12.5
Consumption:				
Libya	3.9	†	†	†
Nigeria	3.6	†	†	†
Algeria	4.6	†	†	†
Egypt	11.5	†	†	†
Angola	†	†	†	†
Gabon	†	†	†	†
South Africa	15.5	41.0*	†	†
TOTAL	63.5	143.1	9.0	12.5

Source: Own Calculations / Various Sources † minimal

Table 142. AFRICA – PRIMARY ENERGY CONSUMPTION 1969–2000

(Unit: Million tonnes oil equivalent)

	OIL		COAL		GAS		HYDRO		TOTAL	
		% of total		% of total		% of total		% of total		% of total
1969	38.8	45.0	40.2	46.6	1.4	1.6	5.9	6.8	86.3	100
1970	42.1	46.0	41.4	45.2	1.5	1.6	6.5	7.1	91.5	100
1971	44.4	48.5	44.4	48.5	1.7	1.9	7.0	7.6	97.5	100
1972	44.7	46.6	42.4	44.2	2.4	21.9	6.4	6.7	96.0	100
1973	49.5	50.4	38.6	39.3	3.2	24.5	6.9	7.0	98.2	100
1974	50.4	49.6	40.4	39.7	3.6	27.1	7.3	7.2	101.7	100
1975	51.5	48.5	43.1	40.6	4.3	24.7	7.3	23.2	106.2	100
1976	55.5	47.7	46.7	40.1	5.1	23.1	9.1	7.8	116.4	100
1977	58.0	47.0	47.3	38.3	6.9		11.2	9.1	123.4	100
1978	61.4	45.5	53.3	39.5	8.3		12.0	8.9	135.0	100
1979	63.5	44.4	58.1	40.6	9.0		12.5	8.7	143.1	100
Yearly change 1969–79	+5.0		+3.8		+20.5		+7.9		+5.2	
Yearly change 1974–79	+4.7		+7.5		+20.0		+11.5		+7.1	
1990	124.1	40	129	42	35	11	16	5	307	100
2000	260.4	43	265	44	56	9	21	3	608	100

Source: Own Calculations. / Various Sources

Table 143. AFRICA GAS CONSUMPTION 1979-2000

(million tonnes oil equivalent)

	1979		1990		2000	
	MTOE	% of Total	MTOE	% of Total	MTOE	% of Total
Nigeria	3.0	33	10	29	17.3	31
Libya	2.5	28	8	23	13.9	25
Algeria	3.0	33	10	29	16.7	30
Others	0.5	5	7	20	8.0	14
AFRICA	9.0	100	35	100	56.0	100

Source: Own Calculations / Various Sources

Table 144 AFRICA OIL CONSUMPTION 1975-2000

(million tons of oil)

	1975		1979		1980		1985		1990		1995		2000	
	MT	% of Total	MT	% of Total	MT	% of Total	MT	% of Total	MT	% of Total	MT	% of Total	MT	% of Total
Nigeria	5.3	13.7	8.6	12.9	9.1	13.7	13.3	14.4	20.7	16.7	32.1	18.7	49.9	19.2
Libya	2.4	6.2	4.5	6.8	4.8	7.2	8.4	9.4	15.6	12.6	28.8	16.8	53.3	20.5
Algeria	3.9	10.0	7.7	12.2	8.2	12.3	13.5	15.1	22.2	17.9	36.6	21.4	60.2	23.1
Others	27.2	70.1	42.7	67.2	44.4	66.8	54.0	60.5	65.6	52.9	73.8	43.1	97.0	37.2
Africa	38.8	100.0	63.5	100.0	66.5	100.0	89.2	100.0	124.1	100.0	171.3	100.0	260.4	100.0

Source: Own Calculations.

Table 145. PER CAPITA CONSUMPTION OF COMMERCIAL PRIMARY ENERGY 1979 – SELECTED COUNTRIES

	Consumption MTOE	Population Millions	Tonnes of Oil Per Capita 1979 Tonnes
Canada	223.4	23.66	9.4
USA	1,895.1	219.63	8.6
Norway	31.3	4.13	7.6
Netherlands	75.7	14.20	5.3
Sweden	44.3	8.33	5.3
Belgium/Luxembourg	52.7	9.94	5.3
West Germany	285.0	62.19	4.6
USSR	1,148.0	269.29	4.3
Denmark	20.3	5.21	3.9
U.K.	220.8	56.24	3.9
Switzerland	24.4	6.47	3.8
France	195.4	54.62	3.6
Austria	26.1	7.58	3.4
Japan	380.7	115.61	3.3
Italy	147.8	58.56	2.5
Greece	20.4	9.17	2.2
Eire	7.0	3.34	2.1
Spain	77.2	37.18	2.1
Yugoslavia	37.9	22.30	1.7
Portugal	11.1	9.06	1.2
China	576.9	941.27	0.6
Turkey	23.1	45.49	0.5

Source: Own Calculations / BP / World Marketing Data and Statistics

Table 146. **PER CAPITA CONSUMPTION OF COMMERCIAL PRIMARY ENERGY 1979–2000**

(Unit: Million tonnes oil equivalent)

	1979			2000		
	Consumption (MTOE)	Population (M)	Tonnes of oil Equivalent Per Capita 1979	Consumption (MTOE)	Population (M)	Tonnes of oil Equivalent Per Capita 2000
N. America	2,121.5	243.29	8.7			
Australasia	84.8	20.65	4.1			
E. Europe	441.8	109.38	4.0			
W. Europe	1,300.5*	414.01	3.1			
Middle East	105.8	84.2	1.3			
Latin America	319.4	342.87	0.9			
S.E. Asia	183.5	442.9	0.4			
Africa	143.1	484.6	0.3			
S. Asia	129.3	873.65	0.1			
W. Hemisphere	2,440.9	586.16	4.2			
E. Hemisphere	4,519.5	3,755.56	1.2			
World (ex USSR, E EUR & China)	4,793.7	3,021.8	1.6			
World	6,960.4	4,341.7	1.6	12,296	6,500	1.9
Developing Countries	881	2,144	0.4	2,447	4,094	0.6
World excluding developing countries	6,079	2,198	2.8	9,849	2,406	4.1

Source: Own Calculations / World Marketing Data and Statistics

Table 147. LESS DEVELOPED COUNTRIES* – ENERGY DEMAND 1979–2000

(Unit: Million tonnes oil equivalent)

	1979		1990		2000		% change 1979–2000	Average growth % per year 1979–2000	Average growth % per year 1969–1979	Average growth % per year 1974–1979
	MTOE	% of total	MTOE	% of total	MTOE	% of total				
Middle East	105.8	12	218	15	485	20	+358	+7.5	+6.0	+1.9
Latin America	319.4	36	486	32	674	27	+111	+3.6	+5.4	+4.6
S.E. Asia	183.5	21	289	19	395	16	+115	+3.7	+7.3	+7.3
Africa	143.1	16	307	20	608	25	+325	+7.1	+5.2	+7.1
S. Asia	129.3	15	195	13	285	12	+120	+3.9	+3.4	+4.5
Total L.D.C.S.	881	100	1,495	100	2,447	100	+178	+5.0	+5.5	+5.1

*includes S. Africa

Source: Euromonitor.

Table 148. LESS DEVELOPED COUNTRIES – CONSUMPTION BY PRIMARY FUEL 1979–2000 (%)

	1979						2000					
	Oil	Coal	Gas	Hydro	Nuclear	Total	Oil	Coal	Gas	Hydro	Nuclear	Total
Middle East	15	-	31	1	-	12	29	-	39	1	2	20
Latin America	42	8	45	45	58	27	36	9	30	73	27	27
S.E. Asia	23	24	8	12	47	21	16	17	14	8	42	16
Africa	13	29	9	16	-	16	21	44	16	10	7	25
S. Asia	7	38	7	13	20	15	5	30	3	8	21	12
Total L.D.C.S.	100	100	100	100	100	100	100	100	100	100	100	100

Source: Own Calculations/ BP / Various Sources

Table 149. WORLD BANK LENDING PROGRAMMES 1981-85 (FISCAL YEARS)

(million 1980 $)

	CURRENT				DESIRABLE			
	Lending Programme		Total Project Cost		Lending Programme		Total Project Cost	
	million $	% of Total	million $	% of Total	million $	% of Total	million $	% of Total
Coal and lignite[1]	840	6	4.270	7	2.000	8	7.350	8
Oil and gas:								
Predevelopment	1.020	8	2.610	4	2.410	10	5.850	6
Oil development[2]	1.755	13	5.900	10	3.320	13	12.150	13
Gas development[3]	1.210	9	3.250	6	2.270	9	5.875	6
Total oil and gas	3.985	30	11.760	20	8.000	32	23.875	26
Refineries	150	1	400	1	1.000	4	3.100	3
Renewables:								
Fuel wood	425	3	850	1	1.100	4	2.200	2
Alcohol	200	1	2.100	4	650	3	4.550	5
Total renewables	625	5	2.950	5	17	-	6.750	7
Electric power	7.590	57	37.950	66	11.000	44	47.450	51
Industrial retrofitting	-	-	0	-	1.250	5	3.825	4
GRAND TOTAL	13.190[4]	100	57.330	100	25.000	100	92.350	100

Source: World Bank

Notes: 1 Includes coal gasification projects
2 Includes heavy oil projects
3 Includes methanol
4 Does not provide for any lending to China

Table 150. **SYNTHETICS CONSUMPTION 1990-2000**

(million tonnes oil equivalent)

	1990	% of Total	2000	% of Total
USA	41	47	223	62
Canada	27	31	59	16
Brazil	6	7	13	4
Venezuela	10	11	46	13
Others	3	3	20	5
WORLD TOTAL	87	100	361	100
World Total Energy	9.506		12.296	

Source: Own Calculations

N.B. Synthetics consumption is included in oil and gas demand projections in earlier tables.

Table 151. WORLD (NON COMMUNIST) ENERGY USE (BY FUEL) 1979

	Oil %	Coal %	Gas %	Hydro %	Nuclear %	TOTAL %
Transport	21	*	*	-	-	21
Industry	8	9	7	-	-	24
Domestic/agriculture	10	2	7	-	-	19
Electricity generation	8	9	4	7	3	30
Non energy use	6	*	*	-	-	6
TOTAL	52	20	18	7	3	100

Source: Own Calculations / Shell

* Less than 1%

Table 152. **WORLD ELECTRICITY USE SHOWING CONVERSION LOSS 1979**

Fuel Input %			Energy Output %		Final Energy Use %	
Oil	26					
Coal	29	Conversion	Transport	*	Industry	37
Gas	13	Loss	Industry	13	Transport	33
Hydro	22	71%	Domestic/agriculture	16	Domestic/agriculture	30
Nuclear	10					
TOTAL	100		TOTAL	29	TOTAL	100

Source: Shell / Own Calculations

* Less than 1%

Table 153. **EXPENDITURE ON RENEWABLE ENERGY MAJOR EEC NATIONS 1974-78**

(million £ sterling - 1975 prices)

	United Kingdom	Federal Republic of Germany	France	Italy
1974	0.48	0.21	3.03	0.95
1975	0.69	3.03	5.09	1.82
1976	0.70	5.07	14.27	2.39
1977	2.83	10.37	18.13	3.50
1978	4.33	15.96	22.26	5.61

Source: States of the European Communities (Doc. No. XII/1000/79-EN) which is in the Library of the House. The conversion to 1975 prices has been made using OECD inflation indices.

Table 154 WORLD WATER POWER* CONSUMPTION 1979–2000

	1979		1990		2000	
	MTOE	% of Total	MTOE	% of Total	MTOE	% of Total
USA	80.1	19	87.0	16	114.0	17
USSR	45.0	11	70.0	13	78.0	11
China	9.0	2	12.0	2	13.6	2
Canada	54.2	13	62.0	12	70.0	10
Western Europe	108.3	26	115.0	22	121.0	18
Japan	19.9	5	45.0	8	57.0	8
Middle East	1.0	0.2	1.2	0.2	1.4	0.2
Africa	12.5	3	16.0	3	21.0	3
Australasia	9.4	2	10.6	2	13.0	2
Latin America	46.4	11	86.0	16	149.9	22
Eastern Europe	5.4	1	7.0	1	9.0	1
South Asia	10.7	3	14.5	3	17.3	2
SE Asia	9.5	2	13.0	2	16.3	2
WORLD	411.5	100	529.3	100	681.5	100
Non Communist	352.4	86	440.3	83	580.9	85

Source: Own Calculations.

* Includes other renewable energy sources.

SUMMARY OF TABLES

Table 68. Imports and Exports 1969-1979 (million tonnes of oil)